DAVID O. MCKAY LIBRARY
P9-CJT-022

BIOCHEMICAL SOCIETY SYMPOSIA

No. 68

FROM PROTEIN FOLDING TO NEW ENZYMES

BIOCHEMICAL SOCIETY SYMPOSIUM No. 68

held at the University of Leeds, April 2000

From Protein Folding to New Enzymes

ORGANIZED AND EDITED BY

A. BERRY AND S.E. RADFORD

PORTLAND PRESS

Published by Portland Press,
59 Portland Place, London W1B 1QW, U.K.
on behalf of The Biochemical Society
Tel: (+44) 20 7580 5530; e-mail: editorial@portlandpress.com
http://www.portlandpress.com

ISBN 1 85578 143 3 ISSN 0067-8964

British Library Cataloguing in Publication Data
A catalogue record for this book is available from the British Library

Typeset by Portland Press Ltd
Printed in Great Britain by Cambridge University Press, Cambridge, UK

Contents

Preface

The second half of the 20th Century saw tremendous advances in our understanding of the chemical basis of life. In this period we have gone from a time when we barely knew how amino acids were strung together to make a protein, or how genetic information was passed from generation to generation, to a time when the sequence of the human genome has been determined and the three-dimensional structures of more than 1000 different proteins, including those of several membrane proteins, have been solved. Moreover, using the now common tools of molecular biology, the amino-acid sequence of individual proteins can be engineered rationally or even evolved for new biological activities. Despite these rapid advances, we are still awaiting a full understanding of the mechanisms by which proteins adopt their folded state, and our ability to predict the structure of proteins or the consequences of mutagenesis on protein structure, stability or activity are still only limited. Research in these areas is dynamic, and exciting new insights are emerging with amazing frequency. In April 2000, the current knowledge in these areas was described and the future directions and potentials debated in the Annual Symposium of the Biochemical Society held at The University of Leeds, U.K.

Leeds University was home to Professor W.T. Astbury, FRS, one of the pioneers of the field of structural molecular biology. Indeed, Astbury first coined the term 'molecular biology' during his time in Leeds, defining it as "...not so much a technique as an approach, an approach from the viewpoint of the so-called basic sciences with the leading idea of searching below large-scale manifestations of classical biology for the corresponding molecular plan. It is concerned particularly with the forms of biological molecules and .... is predominantly three-dimensional and structural — which does not mean, however, that it is merely a refinement of morphology — it must at the same time inquire into genesis and function" [*Nature* (1961) **190**, 1124]. As well as realising the enormous potentials of molecular biology, Astbury was also involved in the early studies of the diffraction patterns of fibrous proteins, particularly those of keratins (see front cover). Using these images, Astbury first recognized alpha to beta transitions in protein structures upon physical perturbation, a phenomenon highly relevant today in the conversion of normally soluble proteins into the toxic confomers of amyloid. It was, therefore, appropriate that a symposium entitled 'From Protein Folding to New Enzymes' that dealt with the structures of proteins, the consequences of misfolding transitions in human disease, and the engineering of structures for new functions, should have been held in Leeds.

We were delighted that many of our distinguished colleagues were able to speak at the meeting, and to share their findings with the readers of this book. The latest views of protein folding mechanisms *in vitro* and insights into some of the structural consequences of protein misfolding are described by Chris Dobson in Chapter 1. The biology of a fascinating yeast prion, as studied using techniques spanning structural biology and yeast genetics to reveal how this protein can cause the inheritance of new traits, is the subject of Chapter 3 by Tricia Serio. Ulrich Hartl and Lila Gierasch (Chapters 4 and 5 respectively) detail some of the amazing feats by which protein folding is chaperoned in cells, while Paula Booth tackles the immensely complex, but fascinating, biological and biophysical problem of how a membrane protein folds in Chapter 2. Here, both the lipid and the protein have to be considered and the reader is taken through the problems and pitfalls of working with folding membrane proteins and the insights that are beginning to emerge from this relatively untapped area.

Some of the most exciting breakthroughs in protein science today are emerging through the combination of experiment and simulations. Our ability to model proteins in action is increasing with remarkable speed, thanks to the enormous increases in computer power available today and the development of algorithms and programs that can model the folding of polypeptides and proteins. These areas are covered here in Chapters 6, 7 and 8, by Valerie Daggett, Andrew Doig and Derek Woolfson, respectively. Valerie Daggett presents the results of molecular dynamics simulations to identify and validate protein folding–unfolding transitions that agree remarkably well with experimental results. Protein design also relies heavily on input from both theory and experiment in cycles of design and redesign. The next two chapters, by Doig and Woolfson respectively, provide insights into advances in this area. Doig summarizes the factors involved in the structure, stability and folding of the α-helix, and Woolfson provides guidelines for the assembly of novel coiled-coil structures. Understanding the fundamental details of the origin of the structure and stability of relatively simple structures such as these will provide the intellectual tools that will permit us to construct new proteins in the future, and to build on the repertoire of proteins that Nature has provided. While rational engineering of enzymes can provide catalysts with altered specificities, these experiments involve often laborious step-wise modifications of existing enzyme activities. Exciting new approaches using directed evolution promise to revolutionize our ability to tailor enzymes for new functions. In Chapter 9, Mike McPherson describes the approaches of DNA shuffling and phage display, and their use in the creation of evolved protease inhibitors that hold enormous promise for the control of nematode pests. The book ends 'with a bang' as Neil Bruce describes how environmental screening has been used to produce new enzymes capable of reducing nitrate esters and their use in the degradation of explosives.

We hope that this book will provide you with an appreciation of the exciting current work in this field and its future potential. Finally, and most importantly, we thank our authors for providing the array of interesting chapters that make up this volume.

Alan Berry
Sheena E. Radford

Abbreviations

AMF	atomic force microscopy
bO	bacterio-opsin
bR	bacteriorhodopsin
CEWC	chicken egg-white cystatin
CI2	chymotrypsin inhibitor 2
CpTI	cow-pea trypsin inhibitor
DHPC	L-α-1,2-dihexanoyl phosphatidylcholine
DMPC	L-α-1,2-dimyristoyl phosphatidylcholine
DOPC	L-α-1,2-dioleoyl phosphatidylcholine
DOPE	L-α-1,2-dioleoyl phosphatidylethanolamine
DPOPC	L-α-1,2-dipalmitoleoyl phosphatidylcholine
DPOPE	L-α-1,2-dipalmitoleoyl phosphatidylethanolamine
GFP	green fluorescent protein
GTN	glycerol trinitrate
GuHCl	guanidine hydrochloride
Hip	Hsc70-interacting protein
Hop	Hsc70/Hsp90-organizing protein
Hsc	heat-shock cognate stress protein
Hsp	heat-shock protein
HSQC	heteronuclear single-quantum coherence
KIH	'knobs-into-holes'
MD	molecular dynamics
NOE	nuclear Overhauser effect
OC-I	oryzacystatin I
OC-IΔD86	OC-I lacking Asp-86
OYE	Old Yellow Enzyme
PC	phosphatidylcholine
PDB	Protein Data Bank
PE	phosphatidylethanolamine
PETN	pentaerythritol tetranitrate
PI3 kinase	phosphoinositide 3-kinase
R-gene	natural resistance gene
RMSD	root mean square deviation
SH3	Src homology 3
SPR	surface plasmon resonance
TNT	trinitrotoluene
TRiC	TCP-1 (T-complex polypeptide 1) ring complex
TSE	transmissible spongiform encephalopathy
VSG	variant surface glycoprotein

Biochem. Soc. Symp. **68**, 1–26
(Printed in Great Britain)

1

Protein folding and its links with human disease

Christopher M. Dobson[1]

Oxford Centre for Molecular Sciences, New Chemistry Laboratory, University of Oxford, South Parks Road, Oxford OX1 3QT, U.K.

Abstract

The ability of proteins to fold to their functional states following synthesis in the intracellular environment is one of the most remarkable features of biology. Substantial progress has recently been made towards understanding the fundamental nature of the mechanism of the folding process. This understanding has been achieved through the development and concerted application of a variety of novel experimental and theoretical approaches to this complex problem. The emerging view of folding is that it is a stochastic process, but one biased by the fact that native-like interactions between residues are, on average, more stable than non-native ones. The sequences of natural proteins have emerged through evolutionary processes such that their unique native states can be found very efficiently even in the complex environment inside a living cell. But under some conditions proteins fail to fold correctly, or to remain correctly folded, in living systems, and this failure can result in a wide range of diseases. One group of diseases, known as amyloidoses, which includes Alzheimer's disease and the transmissible spongiform encephalopathies, involves deposition of aggregated proteins in a variety of tissues. These diseases are particularly intriguing because evidence is accumulating that the formation of the highly organized amyloid aggregates is a generic property of polypeptides, and not simply a feature of the few proteins associated with recognized pathological conditions. That such aggregates are not normally found in properly functional biological systems is again a testament to evolution, in this case of a variety of mechanisms inhibiting their formation. Understanding the nature of such protective mechanisms is a crucial step in the development of strategies to prevent and treat these debilitating diseases.

[1]Present address: Department of Chemistry, University of Cambridge, Lensfield Road, Cambridge CB2 1EW, U.K.

Protein folding and misfolding

A living organism may contain as many as 50000 different types of protein. Following synthesis on the ribosome, each protein molecule must fold into the specific conformational state that is encoded in its sequence in order to be able to carry out its biological function. How this process happens is one of the most fascinating and challenging problems in structural biology [1]. In the cell, folding takes place in a complex and highly crowded environment, and the folding process is aided by a range of auxiliary proteins [2,3]. These proteins include molecular chaperones, whose main role is to protect the incompletely folded polypeptide chain from non-productive interactions, particularly those that result in aggregation, and folding catalysts, whose job is to speed up potentially slow steps in the folding process such as those associated with the isomerization of peptidylprolyl bonds and the formation of disulphide bonds. However, it is evident that the code for folding is contained within the amino-acid sequence of the protein itself because it has been shown that proteins can reach their folded structure *in vitro* in the absence of any auxiliary factors, provided that appropriate conditions can be found [4]. The questions of how the fold is encoded in the sequence, and how the process of folding takes place, are at last beginning to be answered in a credible manner. Progress in this area has come about as a consequence of novel experimental strategies to probe the structural transitions that take place during folding *in vitro*, and of innovative theoretical studies designed to stimulate these transitions [5]. Perhaps of greatest importance has been the fact that these two types of approaches have been brought together in a synergistic manner to advance our fundamental understanding of this highly complex process.

In vivo, the beginning of the folding process is the nascent chain as it emerges from the ribosome. *In vitro*, folding begins from a fully formed polypeptide chain that has been unfolded, usually by addition of a chemical denaturant such as urea. In both cases the polypeptide chain is highly disordered before folding is initiated. In the extreme case, approximated rather well for many proteins in denaturants, the protein is said to be in a 'random coil' state, where only local steric interactions constrain the conformations adopted by the molecule [6,7]. In order to achieve the native state encoded by the fold, the protein molecule has to find its way to this unique conformation rather than one of the countless alternatives. Contemplation of this problem gave rise to the 'Levinthal paradox', which can be stated most simply as the fact that it would take an almost infinite time for even a small polypeptide chain to search all possible conformations to find the correct (lowest energy) structure, yet real proteins fold rapidly, frequently in less than 1 s [8].

The solution to this apparent paradox has emerged recently through the consideration of the energy surfaces or 'landscapes' on which the folding reaction occurs [9–12]. This approach recognizes that folding should not be considered as being analogous to a small molecule reaction, such as breaking and forming a small number of single, strong covalent bonds, but as a biased search on an energy surface that is generally rather flat, i.e. where there are many conformations of similar energy. Such a surface arises because the conformation of a protein is determined by a very large number of relatively weak non-covalent interactions, such as hydrogen bonds and hydrophobic interactions. The bias in folding arises because, on average, the 'correct', i.e. native-like, contacts between residues are more stable than 'incorrect', or non-native, interactions in any protein that can fold successfully [12]. On this statistical or 'new view' of protein folding, a given polypeptide chain within the ensemble of conformations making up the denatured state samples only a small number of conformations during its biased search of conformational space, leading it to increasingly lower energy (Figure 1). This process is, however, a stochastic one, and different members of the initial conformational ensemble form their stabilizing contacts in very different orders. The energy surface is itself determined from the sequence, and the key feature of any sequence that can fold successfully is that the shape of its energy surface is such that all, or at least the large majority, of the folding routes, or trajectories, lead efficiently to the single lowest energy conformation, the native state.

Considerable progress on the calculation of the surfaces appropriate to protein folding has come about as a result of theoretical simulations of the folding process. Because of the complexity of the problem and the limitations of present computer power, these studies have largely used simplified models of proteins. Such models need to be sophisticated enough to encompass key features of protein folding, e.g. there is a Levinthal paradox, but simple enough to allow many simulations to be carried out in a reasonable time [11]. Although molecular dynamics simulations in which all the atoms in the protein are explicitly defined are becoming viable for this purpose, at least for small proteins the most common type of model is a 'lattice model'. A lattice model represents a protein as a string of beads that interact with one another according to a set of simplified potentials (Figure 1). The folding process can then be simulated by using Monte Carlo methods in which moves of the chain are biased towards those that result in lower energies. Such methods have been extremely important in establishing the fundamental principles of folding, i.e. how a protein could fold. In order to discover how an actual sequence does fold, however, experimental data need to be obtained to which the simulations can be related.

Investigating the folding of a protein by experimental methods is challenging, both because of the magnitude and rate of the conformational changes that occur, and because of the extreme heterogeneity of the ensembles of conformations that exist at all but the very last stages in the folding process. There are two main approaches that have been developed to overcome these problems. The first is to use biophysical techniques capable of monitoring the properties of a molecular ensemble as folding progresses [13,14]. Because of the rapidity of the folding process, such methods usually need to be applied in stopped or quenched flow mode and a number of methods need to be used in concert to map different conformational properties of the ensemble. Thus, for example, far ultraviolet circular dichroism (UV CD) can be used to monitor the evolution of secondary structure during a folding reaction, while near UV CD or fluorescence can be used to follow the development of tertiary interactions. An increasing number of techniques is being developed for this purpose on increas-

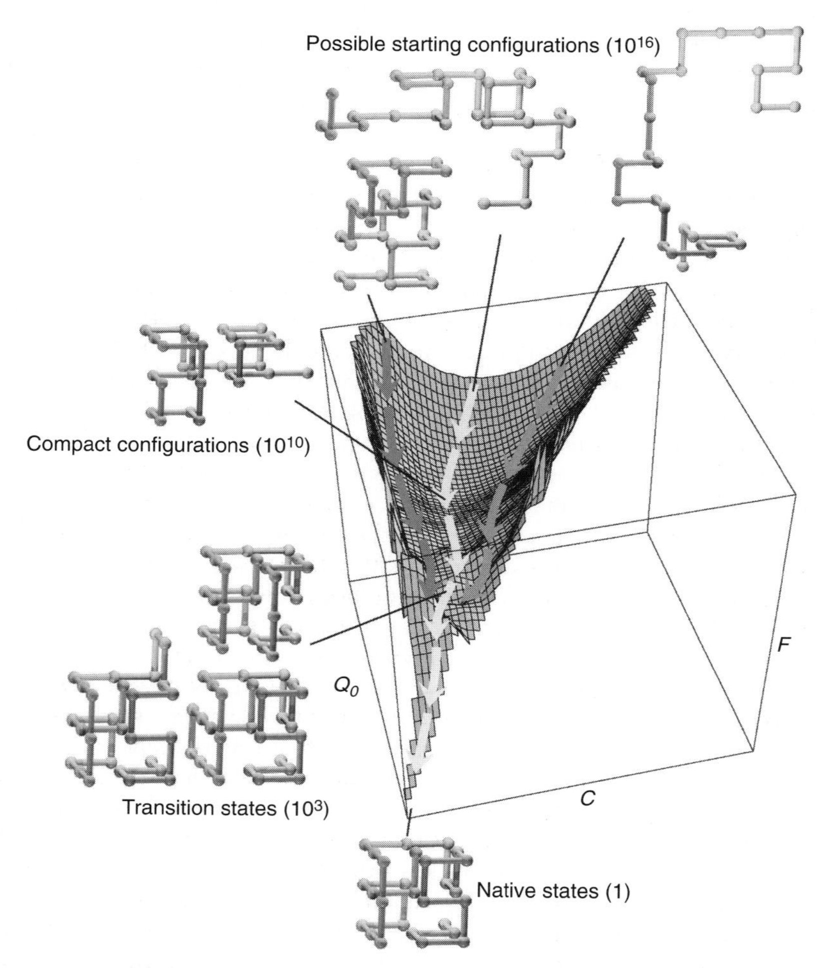

Figure 1 Free-energy (*F*) surface of a 27-mer model protein as a function of the number of native contacts (Q_0) and the total number of (native and non-native) contacts (*C*) obtained by sampling the accessible configuration space using Monte Carlo simulations. The central trajectory shows the average path traced by structures in 1000 independent trials each of which began in a different random conformation. The other two trajectories (right and left) show a range of two standard deviations around the average and are thus expected to include approx. 95% of the trajectories. The structures illustrate the various stages of the reaction. From one of the 10^{16} possible random starting conformations, a folding chain collapses rapidly to a disordered globule. It then makes a slow, non-directed search among the 10^{10} semi-compact conformations for one of the approx. 10^3 transition states that lead rapidly to the unique native state. Reproduced from Dinner, A.R., Sali, A., Smith, L.J., Dobson, C.M. and Karplus, M. (2000) Understanding protein folding via free energy surfaces from theory and experiment. *Trends Biochem. Sci.* **25**, 331–339, with permission from Elsevier Science.

ingly short time-scales [15]. Of particular importance, because of its ability to define structure at the level of individual residues, is nuclear magnetic resonance (NMR) spectroscopy, and significant advances have recently been reported in the application of this technique to study folding [16,17].

The second experimental approach to studying the mechanism of folding is to use the methods of protein engineering [18,19]. This strategy is particularly important because it is able to investigate the nature of the transition state for a folding reaction. The essence of the method is to examine the effects on the folding and unfolding kinetics consequent upon mutation of individual residues in the sequence. This approach has been applied in detail to the study of a series of small proteins that fold with apparent two-state kinetics, and provides dramatic evidence that the rate-determining steps in folding involve the formation of native-like contacts around a small number of key residues. This finding has provided compelling evidence for a 'nucleation–condensation' mechanism of folding whereby the majority of the protein structure forms rapidly once a nucleus of key interactions has formed [20]. As well as providing information about the folding of specific proteins, comparison of the transition states of different proteins is beginning to provide insight into the determinants of the folding process. In particular, the transition states of proteins appear to be similar in proteins of similar native state topology, even if the specific sequences are unrelated, and the rates of folding can be correlated with the pattern of the inter-residue contacts formed within different structures [21–23]. Conclusions of this type strongly support the concept of a common fundamental mechanism for folding, and are highly encouraging that some relatively straightforward principles will emerge to link a sequence to the structure that it encodes.

The folding of small proteins, typically those of less than about 100 residues, appears to be limited by the time required to search for the crucial interactions that are needed to permit rapid progression to the native structure. For larger proteins, however, the folding process is typically more complex and is usually associated with the population of one or more partially folded intermediate states. The reason for this difference is likely to be that larger polypeptide chains have a greater tendency to collapse to a compact state in aqueous solution, even if the contacts between residues in such a state are not those associated with the completely native structure. In such a situation, the rate limiting steps in folding can be the re-organization of inter-residue interactions within a more-or-less disordered collapsed state [11]. Within such a state the barriers to re-organization are likely to be much greater than in extended structures, and indeed may involve the breaking of non-native contacts prior to the formation of stable native ones. This situation can give rise to complex kinetics, and also to distinct heterogeneity in the rate at which different molecules are able to achieve the native structure. In other words, some molecules may form sets of contacts between residues in the initial stages of folding that allow the remainder of the structure to form rapidly. Others, in the statistical process, may form contacts that subsequently generate intermediates that must be substantially re-organized before the complete set of native-like interactions can be formed. Such behaviour is seen both in simulations and in experiments. Indeed, in the case of lysozyme (Figure 2), a protein we have studied in great detail, not only is heterogeneity of folding observed under a variety of circumstances, but there is also evidence that this may arise in part from the nucleation of the folding reaction occurring independently in different regions of the structure [12,24–26].

As the size and complexity of proteins increase, therefore, the folding process becomes more complex. Intermediates with only partially formed structures can be populated and have significant lifetimes. In addition, events that may be termed 'misfolding' may take place during the search for the stable native-like contacts between residues. That such complexities are seen even in the benign environment of a dilute solution of a pure protein suggests that they are even more likely to occur in the crowded environment of the cell. Undoubtedly, molecular chaperones are able to mitigate some of the consequences of this complex behaviour and provide some protection for the incompletely folded chain [3]. However, the idea that proteins can misfold, or fold to intermediates that may undergo undesirable reactions such as aggregation, provides insight into potential problems that can arise during folding even in the best designed environments. Folding and unfolding are also now known to be coupled to many of the key events in the functioning of a biological system, including translocation of proteins across membranes, protein trafficking, secretion of extracellular proteins, and the control and regulation of the cell cycle [27]. Thus, the failure of proteins to fold, or to remain folded under physiological conditions, is likely to cause malfunctions and hence disease [28,29]. Indeed, an increasing number of diseases is now linked to phenomena that can loosely be described as 'misfolding'; a selection of these is given in Table 1.

Protein aggregation and amyloid diseases

Among the diseases in Table 1 are those that are associated with the deposition of proteinaceous aggregates in a variety of organs such as the liver, heart and brain [30–33]. Many of these diseases are described as 'amyloidoses' because the aggregated material stains with dyes such as Congo Red in a manner similar to starch (amylose), the aggregates are referred to as 'amyloid' and the typical fibrous structures (Figure 3) as 'amyloid fibrils'. A list of known amyloid diseases is given in Table 2, along with the major protein component that is associated with the extracellular aggregates in each case [34]. It is evident that these diseases include many of the most debilitating conditions in modern society, particularly those associated with ageing, such as type II diabetes and Alzheimer's disease. Some are familial, some associated with medical treatment (e.g. haemodialysis) or infection (the prion diseases), and some are sporadic (e.g. most forms of Alzheimer's). Some of the diseases (such as the amyloidoses associated with the protein transthyretin) can be found in both sporadic and familial forms. In addition to these diseases, there are others, notably Parkinson's and Huntington's diseases [33,35], that appear to involve very similar aggregates but which are intracellular not extracellular, and are therefore not included in the strict definition of amyloidoses. Many of the deposits include proteins additional to those primarily involved in the fibrillar structures. In addition, there is

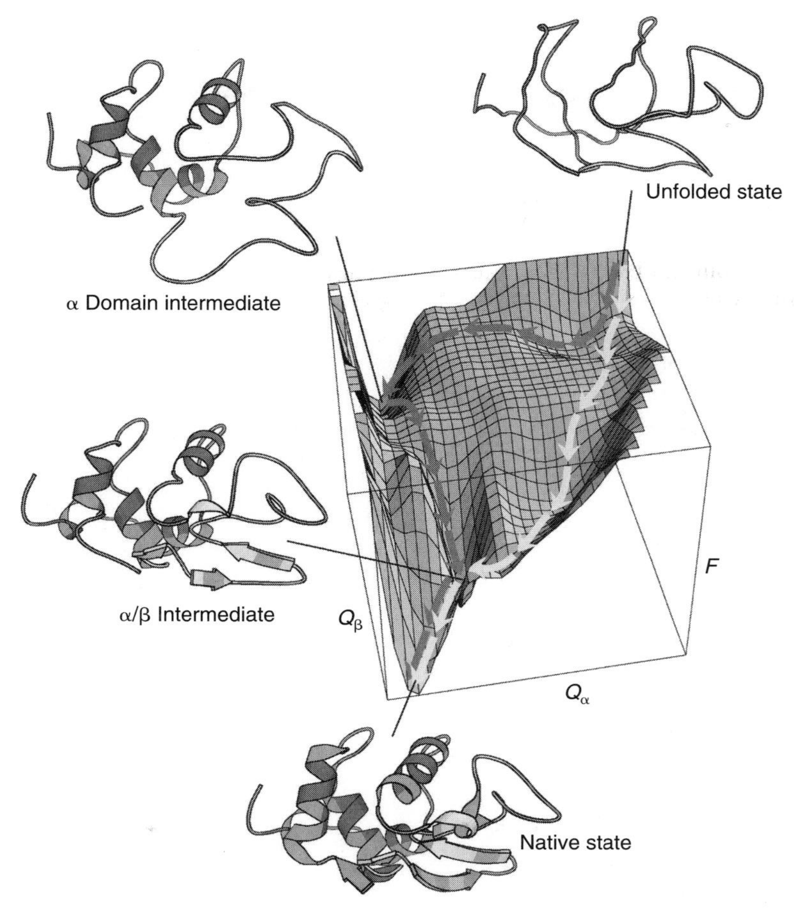

Figure 2 Schematic free-energy (*F*) surface representing features of hen lysozyme (a protein of 129 residues whose structure consists of two domains denoted α and β). Q_α and Q_β are the numbers of native contacts in the α and β domains. The right-hand trajectory represents a 'fast track' in which the α and β domains form concurrently, and populate the intermediate (labelled α/β) only transiently. The left-hand trajectory represents a 'slow track' in which the chain becomes trapped in a long-lived intermediate with persistent structure in only the α domain; further folding from this intermediate involves either a transition over a higher barrier, or partial unfolding to enable the remainder of the folding process to occur along the fast track. Residues whose amide hydrogens are protected from solvent exchange in the native structure (as assessed by NMR) are coloured dark grey (α domain) or white (β domain); all others are lightgrey. In each case, regions indicated to be native-like by monitoring the development of hydrogen exchange protection during kinetic refolding experiments are drawn as ribbon representations of the native secondary structure elements (α-helices and β-sheets). Reproduced from Dinner, A.R., Sali, A., Smith, L.J., Dobson, C.M. and Karplus, M. (2000) Understanding protein folding via free energy surfaces from theory and experiment. *Trends Biochem. Sci.* **25**, 331–339, with permission from Elsevier Science.

Table 1 Representative protein folding diseases. ER, endoplasmic reticulum. Data from [28,30,32,33].

Disease	Protein	Site of folding
Hypercholesterolaemia	Low-density lipoprotein receptor	ER
Cystic fibrosis	Cystic fibrosis transmembrane regulator	ER
Phenylketonuria	Phenylalanine hydroxylase	Cytosol
Huntington's disease	Huntingtin	Cytosol
Marfan syndrome	Fibrillin	ER
Osteogenesis imperfecta	Procollagen	ER
Sickle cell anaemia	Haemoglobin	Cytosol
α1-Antitryspin deficiency	α1-antitryspin	ER
Tay–Sachs disease	β-Hexosaminidase	ER
Scurvy	Collagen	ER
Alzheimer's disease	β-Amyloid/presenilin	ER
Parkinson's disease	α-Synuclein	Cytosol
Creutzfeldt–Jakob disease	Prion protein	ER
Familial amyloidoses	Transthyretin/lysozyme	ER
Retinitis pigmentosa	Rhodopsin	ER
Cataracts	Crystallins	Cytosol
Cancer	p53	Cytosol

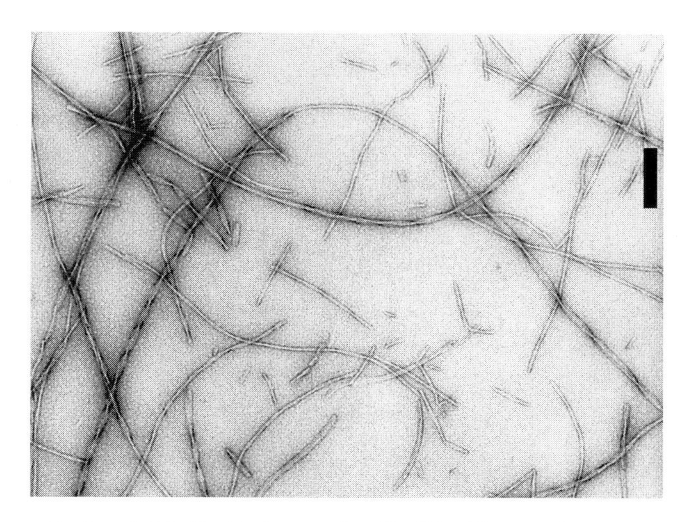

Figure 3 Amyloid fibrils from the Ile-56→Thr variant of human lysozyme viewed by transmission electron microscopy. Scale bar, 200 nm. Reproduced from Morozova-Roche, L.A., Zurdo, J., Spencer, A., Noppe, W., Receveur, V., Archer, D.B., Joniau, M. and Dobson, C.M. (2000) Amyloid fibril formation and seeding by wild type human lysozyme and its disease related mutational varients. *J. Struct. Biol.* **130**, 339–351, with permission from Academic Press.

considerable debate as to whether the fibrillar aggregates themselves give rise to the clinical manifestations of the disease, particularly in the case of brain pathologies such as Alzheimer's and Creutzfeldt–Jakob diseases, and it has been suggested that precursor aggregates of the fibrils may be the primary cause of the diseases, perhaps instigating the destruction of neurons [33]. In systemic amyloidoses, however, it seems likely that the sheer volume of material involved, sometimes kilogram quantities, by itself can disrupt organs such as the liver and the spleen causing them to malfunction [30].

Among the proteins linked with amyloidosis is lysozyme, a protein whose folding we have studied in particular depth. The latter consideration led to the idea that this protein would be an exciting one to choose to try to understand, at the molecular level, the nature of the 'misfolding' transition that converts the protein from a soluble to a fibrillar structure [36]. One of the striking characteristics of amyloid diseases is that the fibrils associated with all of them are very similar in their overall properties and appearance [37]. The fibrils are typically long (often several micrometres), unbranched and approx. 10 nm in diameter (Figure 3). They have a variety of tinctorial properties, notably being stained with Congo Red and exhibiting a green birefringence under polarized light. A range of experiments, particularly X-ray fibre diffraction, indicates that the fibrils have extensive β-sheet character, and that these sheets run perpendicular to the fibril axis to generate what is described as a cross-β structure [37]. This observation is remarkable in view of the fact that the soluble native forms of the proteins associated with these diseases vary

Table 2 Fibril protein components and precursors in amyloid diseases. Data taken from [34].

Clinical syndrome	Fibril component
Alzheimer's disease	Aβ peptide, 1–42, 1–43
Spongiform encephalopathies	Full length prion or fragments
Primary systemic amyloidosis	Intact light chain or fragments
Secondary systemic amyloidosis	76-Residue fragment of amyloid A protein
Familial amyloidotic polyneurotherapy I	Transthyretin variants and fragments
Senile systemic amyloidosis	Wild-type transthyretin and fragments
Hereditary cerebral amyloid angiopathy	Fragment of cystatin-C
Haemodialysis-related amyloidosis	β2-Microglobulin
Familial amyloidotic polyneurotherapy II	Fragments of apolipoprotein A-I
Finnish hereditary amyloidosis	71-Residue fragment of gelsolin
Type II diabetes	Fragment of islet-associated polypeptide
Medullary carcinoma of the thyroid	Fragments of calcitonin
Atrial amyloidosis	Atrial natriuretic factor
Lysozyme amyloidosis	Full length lysozyme variants
Insulin-related amyloid	Full length insulin
Fibrinogen α-chain amyloidosis	Fibrinogen α-chain variants

considerably in nature. Some proteins are large, some small, some are largely helical, some largely sheet. Some are intact in the fibrous form, others are at least partially degraded. Some are cross-linked with disulphide bonds and some are not. This similarity of the fibrillar forms of the proteins prompted the proposal that there are strong similarities in the inherent structure of the amyloid fibrils and in the mechanism by which they are formed [29,37]. Thus the study in depth of the relationship between folding and 'misfolding' of one system could have very general value in understanding this whole class of diseases.

One of the very important observations for this proposal was that the fibrillar forms of many of the disease-related proteins could be generated *in vitro*. In the case of fibrils formed from peptides (often fragments of larger proteins) that are largely unstructured in solution, such fibrils typically form under a wide range of solution conditions. In the case of fibrils formed from intact globular proteins, however, the fibrils typically form under conditions in which the native state is somewhat destabilized [32,38]. Thus, in the case of the two known disease-related human lysozyme variants, fibrils form most readily at low pH or at slightly elevated temperatures [36,39]. Experiments to examine the nature of the amyloidogenic variants (Ile-56→Thr and Asp-67→His) show that the structures of the proteins in their soluble native states are similar to that of the wild-type protein and have no obvious perturbations that could explain their tendency to aggregate [36]. But experiments reveal that the two variants are destabilized relative to the wild-type protein to similar extents, although the origin of this instability is different [40]. Thus, the Ile-56→Thr variant is destabilized largely because its folding rate is reduced, while the Asp-67→His is destabilized largely because it unfolds more rapidly. It therefore appears that the decreased protein stability rather than the altered folding kinetics is a common feature of these two variants. In further experiments it was demonstrated that the lower stability of the native state results in the population of a partially folded state that is very similar to the major (α domain) intermediate population on the folding pathway of the wild-type protein [40]. This finding can be rationalized because the mutations destabilizing the native fold are located in the β domain of the protein, the region that is not highly structured in the predominant intermediate (Figure 4).

This observation suggests a mechanism for the formation of amyloid fibrils from the variant lysozymes, in which the partially folded intermediates aggregate in a first step in the formation of the ordered structure found in the fibrils [36,39,40]. The unfolded region of the intermediate is primarily in the β domain of the protein, suggesting that the aggregation process might be initiated by the intermolecular association of β-strands that are involved in intramolecular interactions in the native structure (Figure 5). This proposition is supported by the observation of the ready formation of amyloid fibrils from a peptide corresponding to part of the β domain of the homologous hen lysozyme [41]. In terms of the energy-landscape model of protein folding, discussed above and illustrated in Figure 2, the mutations destabilize the native state such that, under conditions where the wild-type protein is stable in its native state, the variants may not be. This destabilization can allow the variants to access partially folded states through the fluctuations in structure, inherent

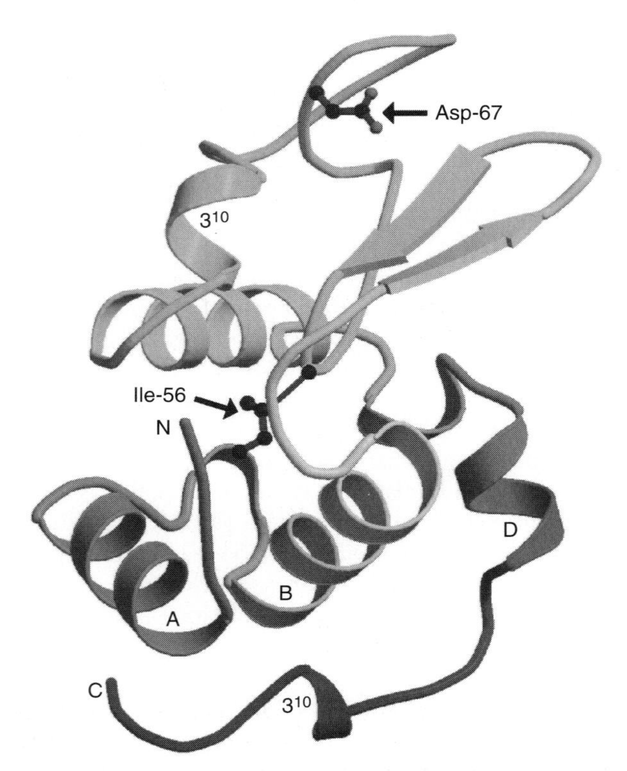

Figure 4 Schematic view of the structure of human lysozyme. The major elements of secondary structure are labelled; the four disulphide bonds are not shown in this representation. The amyloidogenic mutations Ile-56→Thr and Asp-67→His are indicated.

in all proteins, under conditions where the wild-type protein cannot. Calculations based on hydrogen exchange protection suggest that the population of partially folded proteins under physiological conditions is nearly 1000 times greater in the variants than in wild-type lysozyme [40]. This conclusion allows one to speculate that the amyloidogenic variants have sufficient stability to fold efficiently so as to escape the quality control mechanisms in the endoplasmic reticulum and to be secreted into the extracellular environment [29]. However, unlike the wild-type protein, they have insufficient stability to remain in their native states under all conditions to which they are exposed. Moreover, it has been speculated that endosomal compartments where the pH is reduced might be important in the formation of amyloid structures. Under low pH conditions, *in vitro* conversion to amyloid fibrils has been found to be a particularly facile [39]. In addition, *in vitro* experiments have shown that fibril formation is accelerated substantially when solutions are seeded with preformed fibrils (Figure 6). Such a mechanism has been suggested as being responsible for the rapid onset of some amyloidoses, and indeed of the infectivity of the prion diseases [33].

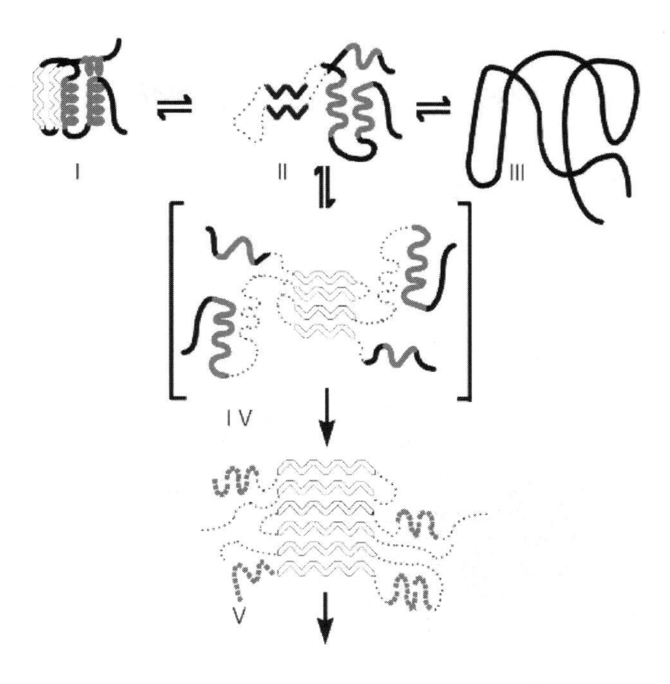

Figure 5 Proposed mechanism for lysozyme amyloid fibril formation. White, β-sheet structure; grey, helical structure; dotted lines, undefined structure. A partially folded form of the protein (II) self associates through the β domain (IV) to initiate fibril formation. This intermediate provides the template for further deposition of protein and for the development of the stable, mainly β-sheet, core structure of the fibril (V). The undefined regions in (V) represent the probability that not all of the polypeptide sequence is involved in the cross-β structure. The nature of this residual structure in (V) is not known, and the figure is not intended to represent any defined secondary structural type. Reproduced from Booth, D.R. et al. (1997) Instability, unfolding and aggregation of human lysozyme variants underlying amyloid fibrillogenesis. *Nature (London)* **385**, 787–793, with permission from Macmillan Publishers Ltd.

The picture emerging from the studies of lysozyme has many features in common with the results of studies of other amyloidogenic proteins, notably those of transthyretin [32]. Of particular importance is the finding that the native state needs to be disrupted to allow fibril formation to occur, and that at least partially unfolded species are accessible under conditions when such fibril formation is rapid. This conclusion is consistent with the observation that most amyloidogenic mutations in the various proteins associated with disease are destabilizing, and that several diseases are associated with fibrils formed by fragments of proteins that are not able to fold into a native-like structure [32]. There are, however, many questions about amyloid formation that remain to be answered. For example, is there any common feature in the group of 20 or so proteins that explains why they form fibrils *in vivo* while other proteins do not

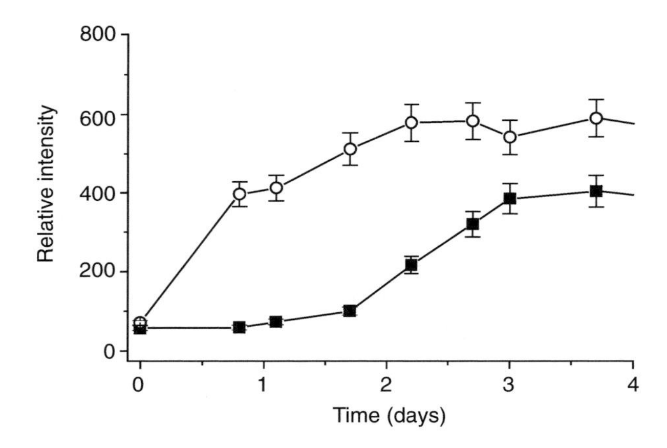

Figure 6 Time dependencies of amyloid fibril formation of the Ile-56→Thr variant of human lysozyme under a particular set of solution conditions. The quantity of amyloid present was monitored through the change in the fluorescence of thioflavin-T associated with its binding to the fibrils. Values are means of four measurements. Filled squares, protein incubated without seeding; open circles, seeding with 2% by volume of a similar situation in which fibrils of Ile-56→Thr lysozyme had previously been allowed to form. Adapted from Morozova-Roche, L.A., Zurdo, J., Spencer, A., Noppe, W., Receveur, V., Archer, D.B., Joniau, M. and Dobson, C.M. (2000) Amyloid fibril formation and seeding by wild type human lysozyme and its disease related mutational varients. *J. Struct. Biol.* **130**, 339–351, with permission from Academic Press.

appear to do so?; and what is the structure of the fibrils that results in them appearing so similar, regardless of the nature of the protein from which they form? A chance observation during studies of the folding of a small SH3 (Src homolgy 3) domain was the stimulant to further work from our laboratory that begins to address both these questions.

The generic nature of amyloid structure

In studies of the conformation of the SH3 domain from bovine PI3 kinase (phosphoinositide 3-kinase) at low pH, when the protein is in a largely unfolded state, it was found that the protein readily formed a viscous gel. Examination of the gel using electron microscopy revealed the presence of large numbers of fibrils that closely resemble those formed from the proteins associated with amyloid diseases [42]. Moreover, the aggregates showed all the other characteristics of amyloid fibrils, and were to all intents and purposes identical to these other structures. This observation prompted us to explore the possibility that similar fibrils could be formed from other proteins by placing them under mildly denaturing conditions that do not immediately result in visible precipitation, and examining the solutions over often prolonged periods of time [38]. For a range of representative proteins with no known connection with any disease we have been able to find conditions under which conversion occurs into fibrils very similar to those associated with amyloid disease (Table 3). We shall refer to these types of structures as 'amyloid fibrils' in future, regardless of whether or not they are associated with disease. The proteins studied included wild-type human lysozyme, which forms fibrils under similar but more destabilizing conditions than the amyloidogenic intermediates [39], and the archetypal globular protein, myoglobin [43]. For the latter it is particularly evident that the protein has undergone a substantial conversion from its soluble α-helical form to the aggregated β-sheet conformation found in the fibrils. Such findings, along with a number of related observations, prompted us to conclude that the ability to form amyloid fibrils is not a characteristic associated wholly or primarily with those proteins found to be associated with amyloidoses, but a property that could be common to many or indeed all proteins under appropriate conditions [29,38,42].

The ability to generate samples from a wide variety of protein sequences provides us with an opportunity to study a much wider range of properties of amyloid fibrils than was previously possible. Indeed, the SH3 fibrils turn out to be particularly well ordered when grown slowly under controlled conditions, and in some cases have an extremely regular helical twist. These features of the fibrils have enabled high-resolution cryo-electron microscopic studies of their structure to be carried out, and a relatively detailed electron density map to be obtained [44]. This study reveals the fibrils to be composed of four 'protofilaments', wound around a hollow core to generate a twisted hollow tube. From the electron density map a tentative model of possible arrangement of the β-sheet structure could be generated (Figure 7). This model, like several earlier proposals, indicates that the structure of the protofilaments is based on hydrogen bonds between the polypeptide main chain [37,44]. As this feature is common to all polypeptides, it explains how the fibrils from different proteins appear so similar, regardless of length and sequence of the polypeptide involved. In contrast to the situation in native proteins, we suggest that the side chains are not a strong influence on the basic structure of the protofilaments. Nevertheless, the manner in which the protofilaments pack together to form mature fibrils may well depend significantly on those parts of the polypeptide

Table 3 Conversion of representative proteins into amyloid fibrils *in vitro*. ADA2h, activation domain of human procarboxypeptidase A2.

Protein	Native structure type	Reference
PI3–SH3 domain	β	[42]
Fn III domain	β	[63]
Acylphosphatase	α/β	[38]
ADA2h	α/β	[59]
Lysozyme	α+β	[39,41]
Cytochrome c_{552}	α	[64]
Apo-myoglobin	α	[43]

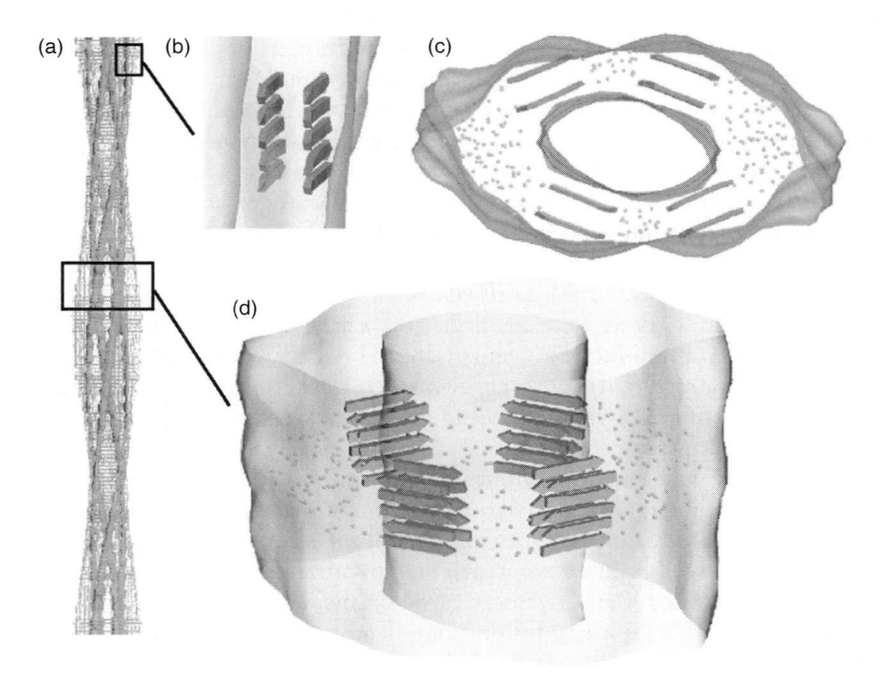

Figure 7 Model of the polypeptide fold in SH3 fibrils (a) Overview of the fibril structure, showing the outer surface as a mesh and the protofilaments as solid surfaces. The ribbon-like protofilaments form the skeleton of the fibril structure. A model for the molecular packing is shown in (b–d), with the electron micrograph map as a transparent rendered surface: (b) side view of a single protofilament; (c) cross section of the fibril; (d) slightly tilted side view of the fibril. β-Sheets derived from the SH3 structure have been fitted into the map, after opening the β-sandwich fold and re-orientating and straightening the strands. The remaining regions of polypeptide sequence are shown as disconnected dots, to indicate the number of residues present but not the conformation. The β-sheets in this model contain a mixture of parallel and antiparallel strands. This particular arrangement is arbitrary and was chosen because it required the least rearrangement of the native β-sheet structure, although there is no evidence that the regions of sheet that are present in the native fold are especially or uniquely present in the fibrils. The β-sheets fit well into the protofilament density, and the loops provide the right amount of mass to generate the rest of the density. Reproduced from Jiménez, J.L., Guijarro, J.L., Orlova, E., Zurdo, J., Dobson, C.M., Sunde, M. and Saibil, H.R. (1999) Cryo-electron microscopy structure of an SH3 amyloid fibril and model of the molecular packing. *EMBO J.* **18**, 815–821, with permission from the European Molecular Biology Organization.

chain that are not involved directly in the close-packed β-strands [45]. Thus, the fibrils from different peptides and proteins are variations on a common theme. One can speculate that the dimensions of the protofilaments, and the lengths of the β-strands within them, may be determined simply by the physio-chemical properties of an unfolded polypeptide chain. Presumably the length of the strands in a regular structure will be a balance between the stabilizing contributions of individual hydrogen bonds and the probability of a turn occurring in the polypeptide chain. It is interesting in this regard that the persistence length of a random coil polypeptide chain is approximately eight to ten residues [7], a range similar to that of the lengths of individual strands in the amyloid structure, and indeed to the typical lengths of strands formed in the native structures of proteins [46].

The proposal that amyloid fibrils are a generic structure of polypeptide chains has stimulated us to suggest that the conformational properties of all proteins should be considered in terms of the multiple states that are accessible to such structures [5,47]. This suggestion is illustrated in Figure 8 in a schematic manner. This diagram suggests that the various fates awaiting a polypeptide chain once it has been synthesized in the cell will depend on the kinetics and thermodynamics of the various equilibria between different possible states. Thus, the normal folding process may pass through partially folded states on the route to the fully native state, but the aggregation of these species will be minimized by the presence of molecular chaperones. In addition, if the protein is able to fold rapidly, any partially folded species will have a short lifetime, reducing the probability of intermolecular interactions occurring. Moreover, once folded, the native state is generally a highly compact structure that conceals the polypeptide main chain within its interior. Such a state is protected from aggregation except through the interactions of surface side chains (as is the case, for example, in protein crystals) and is unable to form the strong intermolecular hydrogen bonds associated with the polypeptide backbone. Provided that the native state is maintained under conditions where it remains folded, aggregation to amyloid fibrils will be resisted by the kinetic barrier associated with unfolding, even if the aggregated state is thermodynamically more stable. Importantly, the co-operative nature of protein structures means that virtually none of the polypeptide chain in individual molecules is locally folded, and that virtually no molecules in an ensemble are globally unfolded, even though native proteins are only marginally stable relative to denatured ones under normal physiological conditions [5,47].

This picture enables us to speculate on the origins of the amyloid diseases from the point of view of the physico-chemical properties of the protein molecules. If the stability or co-operativity of the native state of a protein is reduced, for example by a mutation, the population of non-native states will increase, as discussed above for the amyloidogenic variants of lysozyme. This rise will increase the probability of aggregation, as the concentration of polypeptide chains with at least partial exposure to the external environment will be greater. Whether or not aggregation does occur will depend on the concentration of protein molecules, the intrinsic propensity for a given sequence to aggregate when unfolded, and on the rate of the aggregation process. The fact that formation of ordered amyloid fibrils can be seeded, like the well studied processes of crystallization and gelation, means that once the aggregation process is initiated it often proceeds very rapidly [39,41,48]. In the absence of seeding there can be long 'lag'

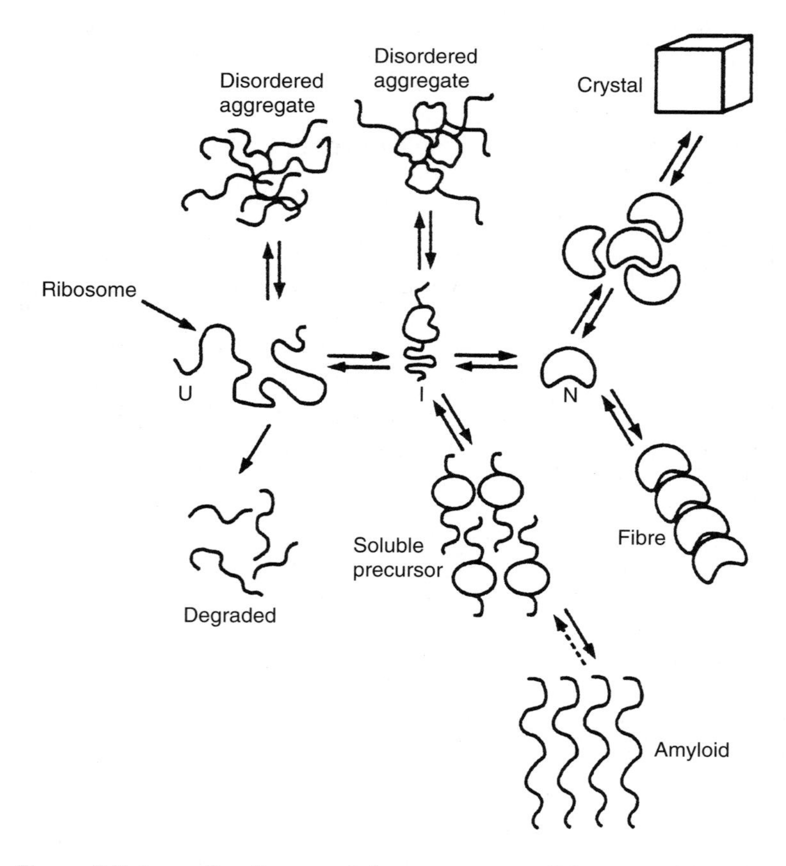

Figure 8 Schematic of some of the states accessible to a polypeptide chain following its biosynthesis. In its monomeric state, the protein is assumed to fold from its highly disordered unfolded state (U) through a partially structured intermediate (I) to a globular native state (N). The native state can form aggregated species, the most ordered of which is a three-dimensional crystal, while preserving its overall structure. The unfolded and partially folded states can form aggregated species that are frequently disordered, but highly ordered amyloid fibrils can form through a nucleation and growth mechanism. Reproduced from Dobson, C.M. (1999) How do we explore the energy landscape for folding? in *Simplicity and Complexity in Proteins and Nucleic Acids* (Fraunfelder, H., Deisenhofer, J. and Wolynes, P.G., eds), pp. 15–37, with permission from Dahlem University Press.

phases before aggregation occurs (Figure 5). This lag can be thought of as arising because the growth of a fibril cannot occur until a 'nucleus' of a small number of aggregated molecules is formed. Such a nucleus can be formed by the local fluctuations in concentration that occur in solution as a result of random molecular motion. When such fluctuations result in a local concentration of molecules above a critical value, the molecules associate with one other to form a species that is sufficiently large to have intrinsic stability, and hence to grow in size by interacting with other molecules in the solution. The act of seeding introduces such nuclei into the solution and hence reduces or abolishes the lag phase.

A general mechanism of amyloid formation

On this view of the aggregation process, the critical step is the unfolding of the native structure. In the case of most proteins, except the smallest ones discussed at the beginning of this article, unfolding under physiological conditions will not generate the type of highly unfolded states seen in high concentrations of denaturant. Instead, the denatured protein will be more stable in a collapsed 'intermediate' state that may well resemble the intermediates observed in the normal folding process [16]. The generic nature of the structure and mechanism of amyloid formation suggests that the nature of the residual structure in such intermediates has little direct importance in dictating the structure of the resulting ordered aggregates, although it may indicate which regions of the protein are most likely to be incorporated into the β-sheet segments of the fibrils. For example, in the case of lysozyme it is likely that the β structure in the fibrils is mainly formed from the β domain that is highly unfolded in the intermediate populated under conditions where amyloid forms [40,41]. The α domain may be incorporated into the fibrils as a disordered but partially structured region connecting the β-strands, as seen in the loop regions of the SH3 structure [44]. The intermediates may be important, however, for another reason in that they are generally much more soluble than highly unfolded polypeptides, enabling the concentrations of non-native species to reach those required for nucleation as described above.

This general view of amyloid formation can readily be extended to include the existence of sporadic as well as familial diseases (i.e. those involving mutations that destabilize the protein or provide the unfolded or partially unfolded polypeptide chain with an increased tendency to aggregate) and infectious diseases (for example, if they are nucleated by seeding with already aggregated species). Such diseases could arise from the loss of the normal control and regulation processes that enable proteins to be maintained in their required states under all conditions in the organism [29,38]. It is, perhaps, particularly likely that such control is lost in ageing, and the majority of the cases of sporadic diseases such as Alzheimer's or type II diabetes are associated with old age. It is significant that in a high proportion of elderly people even wild-type transthyretin, which in its mutant forms is associated with familial amyloidosis, is found as amyloid structures in organs such as the heart [32]. The exact manner in which this happens is unclear, but it could be as the result of statistical factors (comparable with those observed in lag phases *in vitro*), or of changes in the cellular environment, or of the failure of the normal degradation mechanisms for proteins. Interestingly, some of the diseases in this category involve the deposition of peptide fragments, and the process of degradation in compartments such as lysosomes involves conditions such as low pH that serve to unfold proteins prior to the action of proteases. Such mildly denaturing conditions are particularly favourable for the nucleation and growth of amyloid structures.

The idea that amyloid is a generic form of protein structure leads to the question of why it is only associated with disease (although in yeast and some other fungi it has been suggested that amyloid may play a role in normal behaviour (see refs [49,50]). One possible explanation is as follows. The formation of amyloid fibrils is difficult to control, and once formed it is often extremely difficult or impossible to degrade. It is not therefore a favourable material for an efficient living organism to exploit. It seems, therefore, that biological evolution has managed to select some sequences that are able to fold to compact, globular and soluble forms that resist aggregation and conversion to fibrillar structures, at least when protected in a highly stable and controlled environment [29]. These remarkable structures are the native states of proteins that are involved in every process occurring in the cell. The appearance of amyloid deposits in living systems may, therefore, be associated with mutations that destabilize native proteins sufficiently for them to convert into fibrils where the wild-type protein would not, but leave them sufficiently stable to evade the quality control mechanisms in the cell and to function sufficiently normally to allow the organism to develop and reproduce. Amyloid deposits also appear in old age where evolutionary pressure is reduced after the reproductive life span, and in other conditions such as kuru or bovine spongiform encephalopathy (BSE), which are connected with abnormal practices such as ingestion of tissue from other members of the same species [31]. This conceptual picture enables all of these diseases to be rationalized, at least in general terms, on a similar basis.

The inherent propensity for proteins — and indeed other molecular species — to aggregate is therefore one of the primary issues that living systems have to control in order to survive. The development of molecular chaperones is one example of a strategy to minimize the effects of aggregation, and these molecules are now recognized to be crucial in enabling the cell to produce proteins efficiently and to localize them appropriately in the cell [3]. The evolution of sequences with the very special ability to form co-operative globular structures, and of the control mechanisms to maintain these molecules under appropriate conditions, is undoubtedly an even more fundamental mechanism of avoiding aggregation [29]. Just as efficient computers need to be small to enable electrons to move rapidly between components, so living cells need to be small and densely packed to allow molecular diffusion to transfer information efficiently. It is truly a remarkable achievement of evolution to generate in cells solutions of macromolecules at concentrations in excess of 300 $mg \cdot ml^{-1}$, comparable with those of some crystalline solids, without non-specific interactions or aggregation [51–53]. It is interesting to speculate, however, that the material rejected by biology could be useful in modern technology. The structure of amyloid, with organization on a nanometre scale, suggests that it might have important applications as almost infinitely functionizable nanostructures. Indeed, it has already been shown that non-biological groups with novel optical properties can be introduced into peptide-based fibrils, demonstrating the viability of this proposal [54].

The ability of natural proteins to form amyloid structures does not violate the crucial hypothesis that a protein sequence codes for a single fold [4]. The nature of amyloid is that it is not coded for by the sequence, as it is formed as a consequence of interactions involving the common polypeptide backbone of all proteins. Its rate and ease of formation will of course depend on the sequence, both as a consequence of the readiness for different side-chains to pack together within the structure, and as a consequence of the solubility and stability of the sequence in solution [55]. In this sense the formation of amyloid fibrils can be likened to crystallization. Virtually all molecules crystallize (including, for example, individual amino acids), although they differ in their readiness to do so. In general, biology has not used crystalline proteins, except in rare instances such as insulin stored in the pancreas [56]. It is the side-chains, however, that code for the specific fold of globular proteins by their ability to pack together in a unique manner to form compact globular structures. Furthermore, it is not necessary for an alternative protein fold to be encoded in the sequence to allow amyloid fibrils to be formed. We believe that the structures of amyloidogenic intermediates do not define the structure of the resulting amyloid fibrils, as discussed above, but that within the ensemble of conformations that comprise unfolded or partially folded states, structures are accessible that can aggregate through the formation of β-sheets. There is therefore no need for the sequence to code for a specific amyloidogenic intermediate in order for fibrils to develop.

Looking to the future

The generic picture of amyloid structure and the mechanism of its formation provides a conceptual framework for linking together the various pathological conditions associated with deposition of proteinaceous material, and hence to suggest possible general approaches to the prevention or treatment of the whole family of amyloid diseases. From the viewpoint of physical chemistry, the origin of enhanced amyloidogenicity is either a reduced stability of the native state, or the increased tendency to aggregate of any accessible populations of unfolded or partially folded species. Both of these factors can, as we have discussed, result from single amino-acid mutations. One therapeutic strategy is, therefore, to increase the stability of the protein involved. This approach has been explored in detail for transthyretin, where substrate analogues that stabilize the native state have been shown to reduce the tendency for amyloidogenic mutants of the protein to form fibrils [57]. Studies of the small protein acylphosphatase have shown that this is likely to be a general method of reducing the tendency to form amyloid fibrils under conditions where the native state of a protein has low stability [58]. Investigation of acylphosphatase and ADA2h (activation domain of human procarboxypeptidase A2) [55,59] has also indicated that mutations that reduce the aggregation tendency of the denatured state are very effective at inhibiting fibril formation (Figure 9). Indeed, with acylphosphatase, it has been found that rates of aggregation can be reduced by a factor of 1000 by single changes of amino-acid residues at positions in the protein that do not perturb significantly the stability or functional behaviour of the protein [55]. These findings suggest that a strategy such as gene therapy could, in principle, be viable to inhibit an amyloid disease by specific modification of the sequence of the protein whose aggregation is the primary origin of the clinical symptoms.

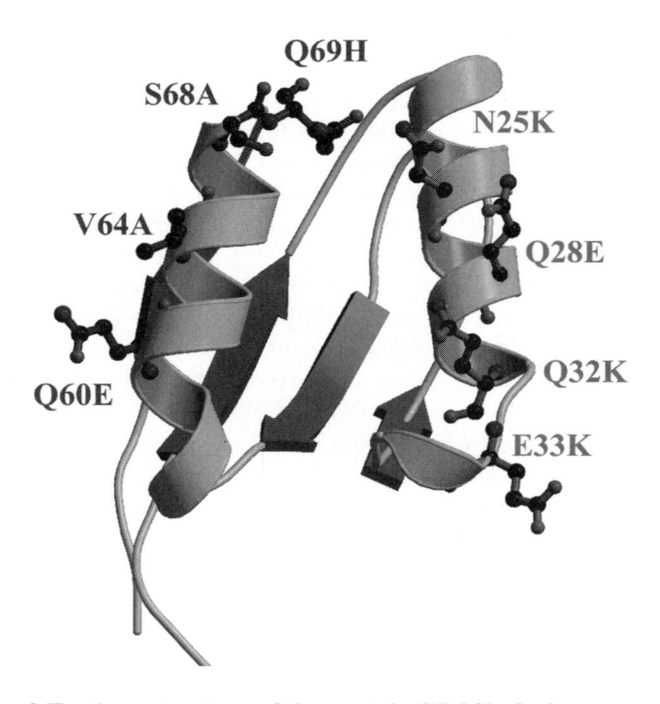

Figure 9 Tertiary structure of the protein ADA2h. Surface mutations in α-helix 1 (N25K, Q28E, Q32K and E33K) and those in α-helix 2 (Q69H, S68A, V64A and Q60E) are shown. Both sets of mutations reduce substantially the propensity of the denatured protein to aggregate, although they have relatively little effect on the structure or stability of the native protein. Reproduced from Villegas, V., Zurdo, J., Filiminov, V.V., Aviles, F.X., Dobson, C.M. and Serrano, L. (2000) Protein engineering as a strategy to avoid formation of amyloid fibrils. *Protein Sci.* **9**, 1700–1708, with permission from Cambridge University Press.

Importantly, it has been demonstrated that liver transplants, which remove the variant proteins associated with amyloidogenesis, can result in remission of disease and gradual disappearance of the fibrils from the body [60].

In addition to these strategies, therapeutic methods have been developed that are designed either to inhibit amyloid formation by blocking fibril growth [61] or to inhibit serum amyloid protein (SAP), a protein that binds to fibrils *in vivo* and protects them from succumbing to the normal degradative processes [62]. Such approaches are very promising, and the enhancement of our knowledge of the mechanism of fibril growth and degradation should help enormously to design effective therapeutic strategies. The fundamental knowledge that can come from the ability to probe the aggregation properties of a range of sequences, rather than just those identified in association with recognized diseases, should enable the role of particular interactions and specific structural motifs in these processes to be explored. This knowledge could result in strategies for drug design that are significantly more general and potentially more effective. Given that many amyloid diseases are associated with old age, the need for novel approaches to therapeutic intervention will be increasingly important in the future. One can speculate that as the average human life span increases, the number of cases of the known diseases is likely to increase substantially, and moreover that novel disorders associated with the aggregation of proteins not so far linked to clinical symptoms might become significant in ageing populations.

As well as the significance of this work in understanding the relationship between folding, misfolding and disease, there are other consequences of the recent findings concerning amyloid formation that have been discussed in this paper. The ability to understand the aggregation behaviour of proteins in more detail, and to develop ways of controlling it, has tremendous possible significance for improving the production and applicability of proteins for commercial purposes. In addition, the ability to control aggregation phenomena has considerable importance in terms of fundamental studies of their properties. In the context of structural biology, for example, control of solubility is of critical importance in inducing efficient crystallization of proteins for X-ray studies, and in generating increased solution concentrations for NMR studies. In addition, the nature of amyloid fibrils suggests that they could form the basis of almost infinitely functional nanomaterials that are biocompatible and highly stable over a wide range of conditions [54]. The development of these various lines of enquiry from observations that were initially somewhat fortuitous [42] illustrates the importance of interdisciplinary research at the interfaces of the physical, biological and medical sciences.

The work described in this article has involved a wide range of students, post-doctoral workers and colleagues in the University of Oxford and elsewhere. The names of many of these people, without whom none of this work would have been possible, are largely to be found in the papers cited in this article. I am particularly indebted to John Ellis for critical reading of this manuscript and for assisting in the preparation of Table 1. The Oxford Centre for Molecular Sciences is supported by the Biotechnology and Biological Sciences Research Council, the Engineering and Physical Sciences Research Council and the Medical Research Council. The research of C.M.D is also supported by the Wellcome Trust.

This chapter is an updated version of an article originally published in *Philos. Trans. R. Soc. London B* (2001) **356**, 133–145. Reproduced with permission from The Royal Society.

References

1. Dobson, C.M. and Fersht, A.R. (eds) (1995) Protein folding. *Philos. Trans. R. Soc. London B* **348**, 1–119
2. Gething, M.-J. and Sambrook, J. (1992) Protein folding in the cell. *Nature (London)* **355**, 33–45
3. Ellis, R.J. and Hartl, F.U. (1999) Principles of protein folding in the cellular environment. *Curr. Opin. Struct. Biol.* **9**, 102–110
4. Anfinsen, C.B. (1973) Principles that govern the folding of protein chains. *Science* **181**, 223–230

5. Dobson, C.M. and Karplus, M. (1999) The fundamentals of protein folding: bringing together theory and experiment. *Curr. Opin. Struct. Biol.* **9**, 92–101
6. Shortle, D. (1996) The denatured state and its role in protein stability. *FASEB J.* **10**, 27–34
7. Smith, L.J., Fiebig, K.M. Schwalbe, H. and Dobson, C.M. (1996) The concept of a random coil. Residual structure in peptides and denatured proteins. *Fold. Des.* **1**, 95–106
8. Karplus, M. (1997) The Levinthal paradox: yesterday and today. *Fold. Des.* **2**, S69–S75
9. Bryngelson, J.D., Onuchic, J.N., Socci, N.D. and Wolynes, P.G. (1995) Funnels, pathways and the energy landscape of protein folding: a synthesis. *Proteins* **21**, 167–195
10. Dill, K.A. and Chan, H.S. (1997) From Levinthal to pathways to funnels. *Nat. Struct. Biol.* **4**, 10–19
11. Dobson, C.M., Sali, A. and Karplus, M. (1998) Protein folding: a perspective from theory and experiment. *Angew. Chem. Int. Ed. Eng.* **37**, 868–893
12. Dinner, A.R., Sali, A., Smith, L.J., Dobson, C.M. and Karplus, M. (2000) Understanding protein folding via free energy surfaces from theory and experiment. *Trends Biochem. Sci.* **25**, 331–339
13. Evans, P.A. and Radford, S.E. (1994) Probing the structure of folding intermediates. *Curr. Opin. Struct. Biol.* **4**, 100
14. Plaxco, K.W. and Dobson, C.M. (1996) Time-resolved biophysical methods in the study of protein folding. *Curr. Opin. Struct. Biol.* **6**, 630–636
15. Callender, R.H., Dyer, R.B. Gilmashin, R. and Woodruff, W.H. (1998) Fast events in protein folding. *Annu. Rev. Phys. Chem.* **49**, 173–202
16. Dobson, C.M. and Hore, P.J. (1998) Kinetic studies of protein folding using NMR spectroscopy. *Nat. Struct. Biol.* **5**, 504–507
17. Dyson, H.J. and Wright, P.E. (1998) Equilibrium NMR studies of unfolded and partially folded proteins. *Nat. Struct. Biol.* **5**, 499–503
18. Matouschek, A., Kellis, J.T., Serrano, L. and Fersht, A.R. (1989) Mapping the transition-state and pathway of protein folding by protein engineering. *Nature (London)* **340**, 122–126
19. Fersht, A.R. (1999) *Structure and Mechanism in Protein Science: a Guide to Enzyme Catalysis and Protein Folding*, W.H. Freeman, New York
20. Fersht, A.R. (1997) Nucleation mechanisms in protein folding. *Curr. Opin. Struct. Biol.* **7**, 3–9
21. Plaxco, K.W., Simons, K.T. and Baker, D. (1998) Contact order, transition state placement and the refolding rates of single domain proteins. *J. Mol. Biol.* **277**, 985–994
22. Chiti, F., Taddei, N., White, P.M. Bucciantini, M., Magherini, F., Stefani, M. and Dobson, C.M. (1999) Mutational analysis of acylphosphatase suggests the importance of topology and contact order in protein folding. *Nat. Struct. Biol.* **6**, 1005–1009
23. Baker, D. (2000) A surprising simplicity to protein folding. *Nature (London)* **405**, 39–42
24. Dobson, C.M., Evans. P.A. and Radford, S.E. (1994) Understanding protein folding — the lysozyme story so far. *Trends Biochem Sci.* **19**, 31–37
25. Thirumalai, D. and Klimov, D.K. (1999) Deciphering the time scales and mechanisms of protein folding using minimal off-lattice models. *Curr. Opin. Struct. Biol.* **9**, 197–207
26. Morozova-Roche, L.A., Jones, J.A., Noppe, W. and Dobson, C.M. (1999) Independent nucleation and heterogeneous assembly of structure folding of equine lysozyme. *J. Mol. Biol.* **289**, 1055–1073
27. Radford, S.E. and Dobson, C.M. (1999) From computer simulations to human disease: emerging themes in protein folding. *Cell* **97**, 291–298
28. Thomas, P.J., Qu, B.H. and Pederson, P.L. (1995) Defective protein folding as a basis of human disease. *Trends Biochem. Sci.* **20**, 456–459
29. Dobson, C.M. (1999) Protein misfolding, evolution and disease. *Trends Biochem. Sci.* **24**, 329–332
30. Tan, S.Y. and Pepys, M.B. (1994) Amyloidosis. *Histopathology* **25**, 403–414
31. Prusiner, S. (1997) Prion diseases and the BSE crisis. *Science* **278**, 245–251

32. Kelly, J.W. (1998) The alternative conformations of amyloid proteins and their multi-step assembly pathways. *Curr. Opin. Struct. Biol.* **8**, 101–106
33. Lansbury, P.T. (1999) Evolution of amyloid: what normal protein folding may tell us about fibrillogenesis and disease. *Proc. Natl. Acad. Sci. U.S.A.* **96**, 3342–3344
34. Sunde, M., Serpell, L.C., Bartlam, M., Fraser, P.E., Pepys, M.B. and Blake, C.C.F. (1997) Common core structure of amyloid fibrils by synchroton X-ray diffractions. *J. Mol. Biol.* **273**, 729–739
35. Perutz, M.F. (1999) Glutamine repeats and neuro-degenerative disease: molecular aspects. *Trends. Biochem. Sci.* **24**, 58–63
36. Booth, D.R., Sunde, M., Bellotti, V., Robinson, C.V., Hutchinson, W.L., Fraser, P.E., Hawkins, P.N., Dobson, C.M., Radford, S.E., Blake, C.C. and Pepys, M.B. (1997) Instability, unfolding and aggregation of human lysozome variants underlying amyloid fibrillogenesis. *Nature (London)* **385**, 787–793
37. Sunde, M. and Blake, C.C.F. (1997) The structure of amyloid fibrils by electron microscopy and X-ray diffraction. *Adv. Protein Chem.* **50**, 123–159
38. Chiti, F., Webster, P., Taddei, N., Clark, A., Stefani, M., Ramponi, G. and Dobson, C.M. (1999) Designing conditions for *in vitro* formation of amyloid protofilaments and fibrils. *Proc. Natl. Acad. Sci. U.S.A.* **96**, 3590–3594
39. Morozova-Roche, L.A., Zurdo, J., Spencer, A., Noppe, W., Receveur, V., Archer, D.B., Joniau, M. and Dobson, C.M. (2000) Amyloid fibril formation and seeding by wild type human lysozyme and its disease related mutational variants. *J. Struct. Biol.* **130**, 339–351
40. Canet, D., Sunde, M., Last, A.M., Miranker, A., Spencer, A., Robinson, C.V. and Dobson, C.M. (1999) Mechanistic studies of the folding of human lysozyme and the origin of amyloidogenic behaviour in its disease related variants. *Biochemistry* **38**, 6419–6427
41. Krebs, M.R.H., Wilkins, D.K., Chung, E.W., Pitkeathly, M.C., Chamberlain, A., Zurdo, J., Robinson, C.V. and Dobson, C.M. (2000) Formation and seeding of amyloid fibrils from wild-type hen lysozome and a peptide fragment form the β-domain. *J. Mol. Biol.* **300**, 541–549
42. Guijarro, J.I., Sunde, M., Jones, J.A., Campbell, I.D. and Dobson, C.M. (1998) Amyloid fibril formation by an SH3 domain. *Proc. Natl. Acad. Sci. U.S.A.* **95**, 4224–4228
43. Fändrich, M., Fletcher, M.A. and Dobson, C.M. (2001) Amyloid fibrils from muscle myoglobin. *Nature (London)* **410**, 165–166
44. Jiménez, J.L., Guijarro, J.L., Orlova, E., Zurdo, J., Dobson, C.M., Sunde, M. and Saibil, H.R. (1999) Cryo-electron microscopy structure of an SH3 amyloid fibril and model of the molecular packing. *EMBO J.* **18**, 815–821
45. Chamberlain, A.K., MacPhee, C.E., Zurdo, J., Morozova-Roche, L.A., Hill, H.A.O., Dobson, C.M. and Davis, J.J. (2000) The ultrastructural organisation of amyloid fibrils by atomic force microscopy. *Biophys. J.* **79**, 3282–3293
46. Branden, C. and Tooze, J. (1999) *Introduction to Protein Structure,* 2nd edn, Garland Publishing, New York
47. Dobson, C.M. (1999) How do we explore the energy landscape for folding? In *Simplicity and Complexity in Proteins and Nucleic Acids* (Fraunfelder, H., Deisenhofer, J. and Wolynes, P.G., eds), pp. 15–37, Dahlem University Press, Berlin
48. Harper, J.D. and Lansbury, P.T. (1997) Models of amyloid seeding in Alzheimer's disease and scrapie: mechanistic truths and physiological consequences of the time-dependent solubility of amyloid proteins. *Annu. Rev. Biochem.* **66**, 385–407
49. Linquist, S. (1997) Mad cows meet *psi*-chotic yeast: the expansion of the prion hypothesis. *Cell* **89**, 495–498
50. Wickner, R.B., Edskes, H.K., Maddelein, M.L., Taylor, K.L. and Moriyama, H. (1999) Prions of yeast and fungi — proteins as genetic material. *J. Biol. Chem.* **274**, 555–558
51. Luby-Phelps, K. (1994) Physical properties of cytoplasm. *Curr. Biol.* **6**, 3–9

52. Minton, A.P. (2000) Implications of macromolecular crowding for protein assembly. *Curr. Opin. Struct. Biol.* **10**, 34–39
53. Ellis, R.J. (2000) Chaperone substrates inside the cell. *Trends Biochem. Sci.* **25**, 210–212
54. MacPhee, C.E. and Dobson, C.M. (2000) Formation of mixed fibrils reveals the generic nature and potential application of protein amyloid structures. *J. Am. Chem. Soc.* **122**, 12707–12713
55. Chiti, F., Taddei, N., Baroni, F., Capanni, C., Stefani, M., Ramponi, G. and Dobson, C.M. (2001) Kinetic partitioning of protein-folding and aggregational (submitted)
56. Dodson, G. and Steiner, D. (1998) The role of assembly in insulin's biosynthesis. *Curr. Opin. Struct. Biol.* **8**, 189–194
57. Klabunde, T., Petrassi, H.M., Oza, V.B., Raman, P., Kelly, J.W. and Sacahetini, J.C. (2000) Rational design of potent human transthyretin amyloid disease inhibitors. *Nat. Struct. Biol.* **7**, 312–321
58. Chiti, F., Taddei, N., Bucciantini, M., White, P., Ramponi, G. and Dobson, C.M. (2000) Mutational analysis of the propensity for amyloid formation by a globular protein. *EMBO J.* **19**, 1441–1449
59. Villegas, V., Zurdo, J., Filiminov, V.V., Aviles, F.X., Dobson, C.M. and Serrano, L. (2000) Protein engineering as a strategy to avoid formation of amyloid fibrils. *Protein Sci.* **9**, 1700–1708
60. Stangou, A.J., Hawkins, P.N., Booth, D.R., O'Grady, J., Jewitt, D., Rela, M., Pepys, M.B. and Heaton, N.D. (1999) Liver transplantation for non-Met30 TTR associated familial amyloid polyneurotherapy. *Hepatology* **30**, 576
61. Soto, C., Sigurdsson, E.M., Morelli, L., Kumar, R.A., Castano, E.M. and Frangione, B. (1998) β-Sheet breaker peptides inhibit fibrillogenesis in a rat brain model of amyloidosis: implications for Alzheimer's therapy. *Nat. Med.* **4**, 822–826
62. Gillmore, J.D., Hawkins, P.N. and Pepys, M.B. (1997) Amyloidosis: a review of recent diagnostic and therapeutic developments. *Br. J. Haematol.* **99**, 245–256
63. Litvinovich, S.V., Brew, S.A., Aota, S., Haudenschild, S.K.C. and Ingham, K.C. (1998) Formation of amyloid-like fibrils by self-association of a partially unfolded fibronectin type III module. *J. Mol. Biol.* **280**, 245–258
64. Pertinhez, T.A., Bouchard, M., Tomlinson, E.J., Wain, R., Dobson, C.M., Ferguson, S.J. and Smith, L.J. (2001) Amyloid fibril formation by a helical cytochrome. *FEBS Lett.* **495**, 184–186

Biochem. Soc. Symp. **68**, 27–33
(Printed in Great Britain)

2

Manipulating the folding of membrane proteins: using the bilayer to our advantage

Paula J. Booth*[1], A. Rachael Curran*[2], Richard H. Templer†, Hui Lu*[3] and Wim Meijberg*[4]

*Department of Biochemistry, Imperial College of Science, Technology & Medicine, London SW7 2AY, U.K., and † Department of Chemistry, Imperial College of Science, Technology & Medicine, London SW7 2AY, U.K.

Abstract

The folding mechanisms of integral membrane proteins have largely eluded detailed study. This is owing to the inherent difficulties in folding these hydrophobic proteins *in vitro*, which, in turn, reflects the often apparently insurmountable problem of mimicking the natural membrane bilayer with lipid or detergent mixtures. There is, however, a large body of information on lipid properties and, in particular, on phosphatidylcholine and phosphatidylethanolamine lipids, which are common to many biological membranes. We have exploited this knowledge to develop efficient *in vitro* lipid-bilayer folding systems for the membrane protein, bacteriorhodopsin. Furthermore, we have shown that a rate-limiting apoprotein folding step and the overall folding efficiency appear to be controlled by particular properties of the lipid bilayer. The properties of interest are the stored curvature elastic energy within the bilayer, and the lateral pressure that the lipid chains exert on the their neighbouring folding proteins. These are generic properties of the bilayer that can be achieved with simple mixtures of biological lipids, and are not specific to the lipids studied here. These bilayer properties also seem to be important in modulating the function of several membrane proteins, as well as the

[1]To whom correspondence should be addressed. Present address: Department of Biochemistry, University of Bristol, Bristol BS8 1TD, U.K.

[2]Present address: Department of Molecular Biophysics and Biochemistry, Yale University, New Haven, CT 06520, U.S.A.

[3]Present address: Department of Medicine, Rayne Institute, University College London, London WC1E 6JJ, U.K.

[4]Present address: Biomade Technology, Nijjenborgh 4, 9747AG Groningen, The Netherlands.

function of membranes *in vivo*. Thus, it seems likely that careful manipulations of lipid properties will shed light on the forces that drive membrane protein folding, and will aid the development of bilayer folding systems for other membrane proteins.

Introduction

The folding mechanisms of integral membrane proteins have largely eluded detailed study. Much of this is due to the inherent difficulties in folding these hydrophobic proteins *in vitro*, which in turn reflects the often apparently insurmountable problem of mimicking the natural membrane bilayer with lipid or detergent mixtures. Membrane proteins can be refolded *in vitro* from a fully denatured state into functional, native proteins. Bacteriorhodopsin (bR) was the first protein that was shown to refold spontaneously, without the need for any accessory proteins [1,2]. Thus, as for water-soluble proteins, the primary amino acid sequence contains all the information needed to define the tertiary structure. Inevitably, the solvent plays a critical role, and refolding of membrane proteins must be performed in detergents or lipids that mimic biological membranes. Although, on the one hand, this makes measurements more technically demanding than those in aqueous solution, on the other, the ability to alter the lipid bilayer structure and dynamics in a known manner offers the potential to control folding.

The folding of bR has been studied more than that of any other membrane protein. Not only can aspects of its folding from a denatured state be studied *in vitro* in a variety of detergents and lipids, but several important folding events can also be followed in a native membrane environment [3–7]. Furthermore, nearly every amino acid in the protein has been mutated individually, and the effect on the generation and the function of the folded state investigated. Several crystal structures for bR have been reported, with the highest resolution being 1.55 Å [8–10]. bR is the only protein constituent of the purple membrane of *Halobacteria salinaria*, where it functions as a light-driven proton pump [11]. The protein consists of seven transmembrane α-helices connected by short extramembrane loops. A retinal chromophore is covalently bound within the helix bundle, via a protonated Schiff-base link, to a lysine residue.

Complete denaturation of bR is possible in organic acids, after removal of the native lipids and the retinal cofactor [1]. SDS is, however, a more suitable denaturant for kinetic studies [12]. The SDS-denatured apoprotein state, bacterio-opsin (bO), is denatured to the extent that it cannot bind retinal, but it retains an α-helical structure equivalent to almost four transmembrane helices. The protein refolds spontaneously on diluting the SDS with renaturing mixed detergent, lipid micelles or lipid vesicles containing retinal. This refolding can be initiated by stopped-flow mixing of equal volumes of the denaturing and renaturing micelles or vesicles, thus giving millisecond time resolution for kinetic studies. Refolding yields of about 95% can be readily obtained, for example, in mixed L-α-1,2-dimyristoyl phosphatidylcholine (DMPC)/CHAPS micelles, DMPC/L-α-1,2-dihexanoyl phosphatidylcholine (DHPC) micelles, native lipid vesicles, L-α-1,2-dipalmitoleoyl phosphatidylcholine (DPOPC) vesicles and DMPC vesicles [6,13,14]. Refolding to native protein from a denatured state can be readily assayed by recovery of the characteristic purple absorption band of the bound retinal chromophore, which is indicative of native-like proton pumping ability.

bR folding kinetics

The simplest reaction scheme that accounts for the kinetic data in lipid-based micelles is [6,13]:

$$\mathrm{bO} \rightleftharpoons \mathrm{I_1} \rightleftharpoons \mathrm{I_2} \overset{\mathrm{R}}{\rightleftharpoons} \mathrm{I_R} \longrightarrow \mathrm{bR}$$

where R is retinal, and I_1 and I_2 are intermediates that form before retinal binding. Retinal binds in at least two steps: first, non-covalently to give I_R and, secondly, via its covalent link to Lys-216 to give bR. The scheme does not include branches or parallel pathways that may well exist. I_R, for example, seems to consist of at least two states (see below), one where the retinal absorption band is similar to that of free retinal at about 380 nm (I_{R380}) and another where the retinal band is red-shifted to 440 nm (I_{R440}).

The intermediate I_1 could be an apoprotein folding intermediate, but also seems to reflect a change in the micelle/vesicle structure as a result of stopped-flow mixing. The apoprotein intermediate I_2 is a key component of the folding process. Formation of I_2 is rate-limiting in apoprotein folding and must occur before retinal can bind. As a result, the formation of I_2 gives rise to an observed lag phase in the formation of folded purple bR [12]. The rate of I_2 formation can also be controlled by manipulating particular characteristics of the refolding lipid environment, with the time constant ranging from seconds to minutes (see later). The changes in protein secondary structure have been time-resolved during this stage of folding by far-UV CD [15]. The SDS-denatured bO state has an α-helical content of about four transmembrane helices, whereas the remaining equivalent of three transmembrane helices are disordered. The secondary structure of I_2 is native-like and corresponds to seven transmembrane helices. About half of the SDS-disordered structure folds to form helices during the 20 s dead-time of these particular far-UV CD experiments, whereas the remaining 30 or so amino acids form helices with a time constant equivalent to that of I_2 formation (i.e. seconds to minutes).

Only one retinal-binding step has been observed in the folding of bR in mixed micelle systems, where retinal binds non-covalently to I_2, probably within some sort of loosely formed binding pocket [16,17]. The exact mechanism by which retinal gains access is unknown. It is also unknown what correct tertiary contacts between the helices are present in the I_2 state, and what controls the specific packing of the helices. There are probably some specific helix–helix contact sites present in the partially folded I_2 state. Retinal binding to I_2 would then allow the helices to pack round the bound retinal to form the folded state. The aqueous loops that connect the helices could also help to provide some specificity and close packing.

Studies of retinal binding to I_2 are complicated by the preceding formation of I_2. However, the binding reaction can be more readily investigated by allowing bO to prefold to a state equivalent to I_2 and then adding retinal [12,16]. There appear to be at least two non-covalent, retinal–protein I_R states, both of which form with the same observed rate of approx. 1.1 s^{-1}. One retinal–protein intermediate (I_{R440}) is observed in transient absorption measurements because the retinal absorption band red-shifts from 380 to 440 nm. However, the formation of I_{R440} cannot by itself account for the observed retinal–protein concentration dependence of the kinetics [17]. It is suggested that another intermediate, I_{R380}, forms and decays in parallel with I_{R440}, with the same observed kinetics, thus giving a reaction scheme where the two parallel paths from I_2 to bR are kinetically indistinguishable:

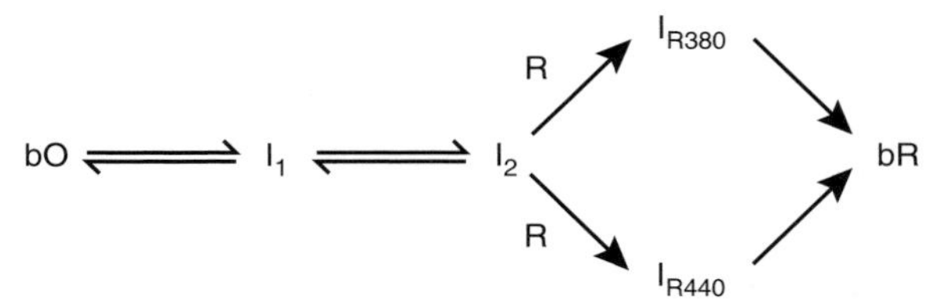

Which of the two routes from I_2 to bR is followed is determined by pH [17]. Optimum refolding, in terms of a maximum refolding yield and overall rate, is observed at pH 6, and the maximum observed concentration of I_{R440} also occurs at pH 6. At pH 8, the refolding yield drops to approx. 80% of that at pH 6, whereas the concentration of I_{R440} drops to 30% of that at pH6. The pH dependence of the two paths from I_2 to bR could result from a distribution of protein conformers in I_2 that have slightly different protonation equilibria of their side-chains and can interconvert due to thermal energy. Binding of retinal to these I_2 conformers would then occur via the same transition state (or distribution of interconverting transition states on a microscopic scale) to form I_R. I_R also contains a similar distribution of protein conformers as I_2; however, the different protonation equilibria of the protein conformations affect the retinal absorption band, and thus show up in this retinal-bound state as I_{R380} and I_{R440}. The retinal absorption band is sensitive to its electrostatic environment. Therefore, an alternative model is possible where the I_2 and I_R states have altered protonation states of individual residues (or bound solvent ions) near retinal, but with no difference in polypeptide conformation. I_{R440} and I_{R380} both decay, with the same observed time constant of a few minutes, to form refolded bR with a 560 nm absorption band in which the Schiff base bond has formed and retinal is covalently bound to Lys-216. This bR state contains all-*trans* retinal within its binding pocket. The retinal then isomerizes at a much slower rate (time constant about 1 h) to a state equivalent to the dark-adapted state of bR that contains a mixture of all-*trans*- and 13-*cis*-retinal [17].

Lipid folding systems

Biological membranes contain a large diversity of lipids, with most membranes containing a mixture of bilayer- and non-bilayer-forming lipids. This seems to have a generic, functional importance by allowing fine tuning of certain bilayer properties that seem to be under homoeostatic control in living membranes, and is vital for the correct function of several of the constituent membrane proteins [18,19]. The presence of the non-bilayer lipids increases the desire of each monolayer of the bilayer to move towards water. However, the monolayers tend to bend in opposite directions, which they cannot do in a bilayer structure. As a result there is a build up of a curvature elastic stress within the membrane and a redistribution of the intermolecular lipid forces, as the monolayers are forced to lie flat, back-to-back, in the bilayer. This is accompanied by an increase in the lateral pressure at the centre of the bilayer, as a result of the increase in the number of collisions between the lipid hydrocarbon chains. There is mounting evidence that these intermolecular forces within lipid bilayers play key roles, both *in vitro* and *in vivo*. The novel crystallization method reported for bR [20], which has led to near-atomic resolution [10,21], seems to be dependent on the manipulation of the lipid intermolecular forces. The introduction of non-bilayer lipids into bilayers has also been shown to modulate the function of several membrane-bound proteins, including alamethicin, cytidyl transferase, rhodopsin, Ca^{2+} ATPase and lactose permease [22–26].

We are currently investigating how these changes in lipid-chain lateral pressures affect the folding of bR. Our aims are two-fold: first, we wish to investigate the molecular origin of forces that control folding *in vitro*, and possibly also *in vivo*; secondly, we aim to develop efficient folding systems for membrane proteins. We have shown that an increase in the lateral pressure within the bilayer may be responsible for the slowing of a rate-limiting folding step for bR [13]. This study used a two-component lipid mixture consisting of lipids with phosphatidylcholine (PC) headgroups but different chain lengths: six-carbon DHPC and 14-carbon DMPC. Increasing the amount of the longer-chain 14-carbon lipid increases the lateral pressure in the chain region and slows the rate-limiting folding step of bR (i.e. formation of I_2). Thus it appears possible to control the rate of protein folding. These DMPC/DHPC mixtures form micelle structures in solution and are thus amenable to the optical methods (fluorescence, absorption and CD) used to study the folding kinetics. However, the exact nature of these mixed micelles at different lipid compositions is unknown and there are additional, unknown constraints on the lipid dynamics and pressures in the micelles, as opposed to a bilayer system.

Lipid-bilayer vesicle systems have been developed for the refolding of bR [14]. Two-component lipid mixtures are used where the desire for monolayer curvature and the lateral pressure in the lipid-chain region can be increased by changing either the lipid headgroup or the lipid chain. Changing the headgroup from PC to phosphatidylethanolamine (PE) increases the desire for monolayer curvature towards water, as does incorporating an unsaturated bond into the lipid chains. Several systems have proved successful for bR refolding: L-α-1,2-dioleoyl phosphatidylcholine (DOPC)/L-α-1,2-dioleoyl phosphatidylethanolamine (DOPE) (18-carbon unsaturated chains with a *cis* double bond at position C-9:C-10), DPOPC/L-α-1,2-dipalmitoleoyl phosphatidylethanolamine (DPOPE) (16-carbon unsaturated chains with a *cis* double bond at position C-9:C-10), DMPC/L-α-1,2-dimyristoyl phosphatidylethanolamine (14-carbon saturated chains) and DMPC/DOPC [14]. Increasing the amount of PE in the PC/PE mixtures, or the chain unsaturation in the DMPC/DOPC mixture, increases the lipid-chain lateral pressure. Furthermore, the different chain lengths in the PC/PE mixtures give rise to different lateral pressures, with for example the longer, 18-carbon dioleoyl chains having a greater chain pressure than the 16-carbon dipalmitoleoyl. It is possible to control the overall refolding yield of bR in these lipid systems. All 100% PC systems allow bR to fold to approx. 100% yield. The yield then decreases as the PE content increases, with a greater decrease being observed in the DOPC/DOPE system than in the DPOPC/DPOPE system [14]. This reflects either an inability of the denatured protein to insert into the stressed bilayer, or an effect of the lipids on protein folding within the bilayer and the accumulation of an intermediate involved in the protein's assembly. We are currently investigating how the slowing of the rate-limiting folding step to I_2, which occurs with lipid-chain pressure, correlates with the decrease in folding yield. Time-resolved optical measurements are more difficult in the lipid vesicle systems than in the DMPC/DHPC micelles, because the lipid vesicles scatter light and can change with time during the course of a folding reaction.

Conclusions

Studies of membrane-protein folding mechanisms lag severely behind those of water-soluble proteins and remain a major challenge in protein research. A major difficulty in the study of membrane proteins lies in finding appropriate detergent or lipid systems that maintain the structural and functional integrity of the protein. This is highlighted in refolding studies where appropriate solubilization conditions must be found both to unfold and refold the protein and, at the same time, to prevent aggregation of the often highly hydrophobic proteins. There is, however, a large body of information on lipid properties and, in particular, on PC and PE lipids, which are common to many biological membranes. We have shown how this can be exploited to develop efficient *in vitro* lipid-bilayer folding systems. Furthermore, we have shown that a rate-limiting apoprotein-folding step and the overall folding efficiency appear to be controlled by particular properties of the lipid bilayer. The properties of interest are the stored curvature elastic energy within the bilayer and the lateral pressure that the lipid chains exert on their neighbouring folding proteins. These are generic properties of the bilayer that can be achieved with simple mixtures of many types of biological lipid and are not specific to the lipids studied here. These bilayer properties also seem to be important in modulating the function of several membrane proteins as well as membrane function *in vivo*.

It is also possible to perform detailed biophysical studies on membrane protein folding and to identify intermediate states and folding mechanisms *in vitro*. At least two intermediates are involved in the folding of bR *in vitro*. It has also been possible to identify two parallel folding pathways during the final stages of folding. These paths are kinetically indistinguishable and are determined by pH, indicating that they arise from different protonation states of the protein.

References

1. Huang, K.-S., Bayley, H., Liao, M.-J., London, E. and Khorana, H.G. (1981) J. Biol. Chem. **256**, 3802–3809
2. London, E. and Khorana, H.G. (1982) J. Biol. Chem. **257**, 7003–7011
3. Khorana, H.G. (1988) J. Biol. Chem. **263**, 7439–7442
4. Haupts, U., Tittor, J. and Oesterhelt, D. (1999) Annu. Rev. Biophys. Biomol. Struct. **28**, 367–399
5. Dale, H. and Krebs, M.P. (1999) J. Biol. Chem. **274**, 22693–22698
6. Booth, P.J. (1997) Folding Design **2**, R85–R92
7. Luneberg, J., Widmann, M., Dathe, M. and Marti, T. (1998) J. Biol. Chem. **273**, 28822–28830
8. Henderson, R., Baldwin, J.M., Ceska, T.A., Zemlin, F., Beckmann, E. and Downing, K.H. (1990) J. Mol. Biol. **213**, 899–929
9. Subramaniam, S. (1999) Curr. Opin. Struct. Biol. **9**, 462–468
10. Luecke, H., Schobert, B., Richter, H.-T., Cartailler, J.-P. and Lanyi, J.K. (1999) J. Mol. Biol. **291**, 899–911
11. Oesterhelt, D. and Stoeckenius, W. (1974) Methods Enzymol. **31**, 667–679
12. Booth, P.J., Flitsch, S.L., Stern, L.J., Greenhalgh, D.A., Kim, P.S. and Khorana, H.G. (1995) Nat. Struct. Biol. **2**, 139–143
13. Booth, P.J., Riley, M.L., Flitsch, S.L., Templer, R.H., Farooq, A., Curran, A.R., Chadborn, N. and Wright, P. (1997) Biochemistry **36**, 197–203
14. Curran, A.R., Templer, R.H. and Booth, P.J. (1999) Biochemistry **38**, 9328–9336
15. Riley, M.L., Wallace, B.A., Flitsch, S.L. and Booth, P.J. (1997) Biochemistry **36**, 192–196
16. Booth, P.J., Farooq, A. and Flitsch, S.L. (1996) Biochemistry **35**, 5902–5909
17. Lu, H. and Booth, P.J. (2000) J. Mol. Biol. **299**, 233–243
18. Gruner, S.M. (1985) Proc. Natl. Acad. Sci. U.S.A. **82**, 3665–3669
19. Bloom, M., Evans, E. and Mouritsen, O.G. (1991) Q. Rev. Biophys. **24**, 293–397
20. Landau, E.M. and Rosenbusch, J.P. (1996) Proc. Natl. Acad. Sci. U.S.A. **93**, 14532–14535
21. Balrhali, H., Nollert, P., Royant, A., Menzel, C., Rosenbusch, J.P., Landau, E.M. and Pebay-Peroula, E. (1999) Structure **7**, 909–917
22. Keller, S.L., Bezrukov, S.M., Gruner, S.M., Tate, M.W., Vodyanoy, I. and Parsegian, V.A. (1993) Biophys. J. **65**, 23–27
23. Lewis, J.R. and Cafiso, D.S. (1999) Biochemistry **38**, 5932–5938
24. Chen, C.C. and Wilson, T.H. (1984) J. Biol. Chem. **259**, 10150–10158
25. Brown, M.F. (1994) Chem. Phys. Lipids **73**, 159–180
26. Navarro, J., Toivio-Kinnucan, M. and Racker, E. (1984) Biochemistry **23**, 130–135

Biochem. Soc. Symp. **68**, 35–43
(Printed in Great Britain)

3

Self-perpetuating changes in Sup35 protein conformation as a mechanism of heredity in yeast

Tricia R. Serio*, Anil G. Cashikar†, Anthony S. Kowal†, George J. Sawicki† and Susan L. Lindquist*†[1]

*Department of Molecular Genetics and Cell Biology, The University of Chicago, Chicago, IL 60637, U.S.A., and †Howard Hughes Medical Institute, The University of Chicago, Chicago, IL 60637, U.S.A.

Abstract

Recently, a novel mode of inheritance has been described in the yeast *Saccharomyces cervisiae.* The mechanism is based on the prion hypothesis, which posits that self-perpetuating changes in the conformation of single protein, PrP, underlie the severe neurodegeneration associated with the transmissible spongiform enchephalopathies in mammals. In yeast, two prions, [*URE3*] and [PSI^{+}], have been identified, but these factors confer unique phenotypes rather than disease to the organism. In each case, the prion-associated phenotype has been linked to alternative conformations of the Ure2 and Sup35 proteins. Remarkably, Ure2 and Sup35 proteins existing in the alternative conformations have the unique capacity to transmit this physical state to the newly synthesized protein *in vivo*. Thus, a mechanism exists to ensure replication of the conformational information that underlies protein-only inheritance. We have characterized the mechanism by which Sup35 conformational information is replicated *in vitro*. The assembly of amyloid fibres by a region of Sup35 encompassing the N-terminal 254 amino acids faithfully recapitulates the *in vivo* propagation of [PSI^{+}]. Mutations that alter [PSI^{+}] inheritance *in vivo* change the kinetics of amyloid assembly *in vitro* in a complementary fashion, and lysates from [PSI^{+}] cells, but not [psi^{-}] cells, accelerate assembly *in vitro*. Using this system we propose a mechanism by which the alternative conformation of Sup35 is adopted by an unstructured oilgomeric intermediate at the time of assembly.

[1]To whom correspondence should be addressed.

Background

In 1965, Brian Cox described a modifier of translation termination efficiency, [*PSI*$^+$], in the yeast *Saccharomyces cerevisiae* [1]. The phenotype produced by the [*PSI*$^+$] element was first detected in strains carrying a nonsense mutation in an auxotrophic marker and a nonsense suppressor tRNA that was too weak to restore growth. The presence of [*PSI*$^+$] caused an increase in the efficiency of the same weak tRNA suppressor, resulting in a detectable level of nonsense suppression (stop codon read-through), and the restoration of growth on selective media.

Early analysis of these strains indicated that the [*PSI*$^+$] phenotype (nonsense suppression) is inherited in a non-conventional manner. [*PSI*$^+$] strains convert spontaneously into the [*psi*$^-$] state at a low frequency, a process referred to as [*PSI*$^+$] curing [1]. The frequency of these events can be increased by growth in the presence of agents that are not mutagenic to nucleic acids, such as methanol or guanidine hydrochloride (GuHCl) [2]. Remarkably, [*psi*$^-$] strains also have the capacity to acquire [*PSI*$^+$] *de novo* at a similarly low, but detectable, frequency [3]. This metastability is difficult to reconcile with a nucleic acid determinant.

Genetic analysis indicated that the [*PSI*$^+$] phenotype is a dominant trait that can be transmitted to either mitotic or meiotic progeny through the cytoplasm [1]. In addition, [*PSI*$^+$] can be transmitted to other yeast strains if their cytoplasms are allowed to mix in the absence of nuclear fusion [4,5]. Despite these observations, subsequent experiments failed to link the [*PSI*$^+$] phenotype to any of the known cytoplasmic nucleic acids [6,7].

Sup35 and the prion hypothesis for [*PSI*$^+$]

The first insights into the nature of the [*PSI*$^+$] determinant were provided by a series of experiments that established a relationship between [*PSI*$^+$] and a nuclear gene known as *SUP35*. *SUP35* had previously been implicated in translation termination [8,9] and has subsequently been shown to be the eukaryotic release factor 3 (eRF3), a component of the translation-termination complex [10,11]. Extra copies of the *SUP35* gene induced a nonsense suppressor phenotype in yeast strains [12], and this phenotype persisted even after the *SUP35* copy number had returned to wild-type levels [13]. Notably, this suppression phenotype was dependent upon expression of the Sup35 protein [14] and, like [*PSI*$^+$], could be cured by GuHCl [13].

In 1994, Reed Wickner [15] proposed a model for the link between *SUP35* and [*PSI*$^+$]. This idea, the yeast prion hypothesis, suggests that the [*PSI*$^+$] determinant is a self-perpetuating alternative physical state of the Sup35 protein, that is, a protein state that can act as a genetic element. This concept is based upon the mammalian prion hypothesis that had previously been proposed to explain the etiology of the transmissible spongiform encephalopathies (TSEs) [16,17]. These neurodegenerative disorders are unique in that they can be transmitted as both genetic and infectious diseases [18]. The infectious agent is widely believed to be a protein with an altered self-perpetuating conformation.

Several criteria accompany the hypothesis that proteins can act as genetic elements in yeast [15]. First, the determinant, a prion protein, must have the capacity to exist in at least two stable but physically distinct states. Secondly, these unique states must be linked to distinct phenotypes. Thirdly, a mechanism must exist for replicating the prion state. Molecular-genetic and cell-biological experiments have provided support for each of these points in the case of Sup35.

In [*PSI*$^+$] strains, Sup35 protein is present in the cytoplasm in large complexes that have an increased resistance to proteolysis; in [*psi*$^-$] strains, Sup35 protein is soluble and sensitive to proteolysis [19,20]. These observations formed the basis of the first molecular model for the [*PSI*$^+$] phenotype [19,20]. The complexed Sup35 protein in [*PSI*$^+$] strains is precluded from fulfilling its function in translation termination, and nonsense codons are read through at an increased frequency. Sup35 protein remains soluble in [*psi*$^-$] strains and thus can act at the ribosome to ensure proper termination.

If the alternative physical state of Sup35 protein is indeed the determinant of the suppression phenotype, conversion between the [*PSI*$^+$] and [*psi*$^-$] states should lead to a corresponding change in the physical state of Sup35 protein: this is the case. If [*PSI*$^+$] strains are cured by treatment with GuHCl, Sup35 protein changes from the insoluble to the soluble form, as assayed by differential

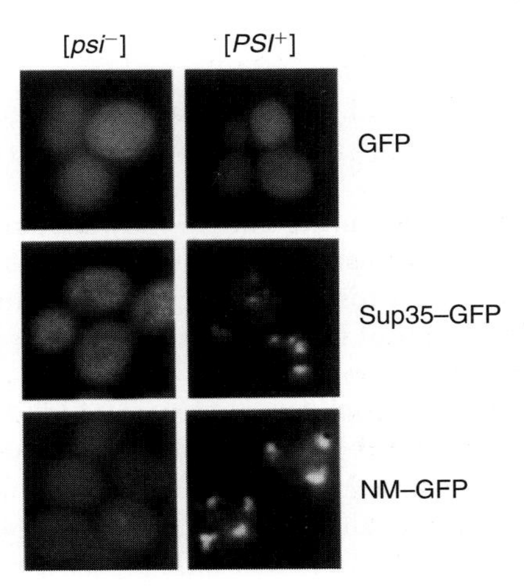

Figure 1 Sup35 adopts distinct physical states corresponding to the [*psi*$^-$] and [*PSI*$^+$] phenotypes. The GFP alone or fusions of GFP with either full-length Sup35p (Sup35–GFP) or the NM region (NM–GFP) were expressed from the copper-inducible *CUP1* promoter in [*psi*$^-$] and [*PSI*$^+$] yeast strains. Expression was induced by the addition of 50 μM $CuSO_4$ to the media for 4 h at 30°C. Fluorescence was visualized under blue light at a ×40 magnification. GFP alone is diffusely distributed throughout the cytoplasm in both [*psi*$^-$] and [*PSI*$^+$] cells. Fluorescence from either Sup35–GFP or NM–GFP is diffuse in [*psi*$^-$] cells, but coalesces into discrete cytoplasmic foci in [*PSI*$^+$] cells.

centrifugation; and, if [*PSI*$^{+}$] is re-introduced by extra-copy *SUP35*, Sup35 changes from the soluble to the insoluble form [19,20]. This correlation is demonstrated most strikingly by the relationship between *SUP35* and the molecular chaperone Hsp104 (heat-shock protein of 104 kDa) [19–21]. [*PSI*$^{+}$] can be cured by either the deletion or transient overexpression of Hsp104 and the change in phenotype is associated with a corresponding change in Sup35 protein from the insoluble to the soluble form. Since the only known function of Hsp104 is to alter the physical state of substrate proteins [22,23], this observation provides strong support for the yeast prion hypothesis in the case of [*PSI*$^{+}$].

As is the case for nucleic acid determinants, a mechanism must exist to ensure replication of the prion state of a protein if the associated phenotype is to persist and be transmitted to progeny. Replication of the prion state for Sup35 was visualized in a series of experiments monitoring the pattern of fluorescence from a protein fusion between Sup35 and the green fluorescent protein (GFP) (Figure 1) [19]. When expressed in yeast, GFP is diffusely distributed throughout the cytoplasm in both [*PSI*$^{+}$] and [*psi*$^{-}$] strains. Sup35–GFP is also diffusely distributed throughout the cytoplasm in [*psi*$^{-}$] strains; however, it rapidly coalesces into discrete foci as soon as it can be visualized in [*PSI*$^{+}$] strains. Immunofluorescence studies using haemagglutinin-tagged Sup35 protein indicate that as Sup35–GFP coalesces into discrete foci, it localizes to pre-existing foci of endogenous Sup35 protein (J. Liu and S. Lindquist, unpublished work). These observations suggest that newly synthesized Sup35–GFP is captured by pre-existing Sup35 protein complexes present in the [*PSI*$^{+}$] cytoplasm.

Several lines of evidence indicate that Sup35–GFP foci provide a direct visualization of the [*PSI*$^{+}$] determinant [19].

(i) If Sup35–GFP is expressed in [*psi*$^{-}$] strains for longer periods, foci begin to appear at the same time as when [*PSI*$^{+}$] colonies can be isolated from the culture.

(ii) In sequential rounds of curing and re-introduction of [*PSI*$^{+}$], fluorescence from Sup35–GFP changes from punctate to diffuse to punctate.

(iii) When Sup35–GFP is expressed in strains carrying mutations in the nucleotide-binding sites of Hsp104, a mixed pattern of fluorescence, both punctate and diffuse, is observed.

This pattern explains the unusual, 'cryptic' state of [*PSI*$^{+}$] in these cells: the cells do not manifest the [*PSI*$^{+}$] phenotype themselves, yet they reproducibly regain that phenotype when the Hsp104 mutation is lost. Translation normally terminates at stop codons because there is sufficient soluble Sup35 protein, owing to the compromised Hsp104 function. Nonsense suppression reappears when the mutant alleles of Hsp104 are segregated because the insoluble Sup 35 that remains can continue to propagate [*PSI*$^{+}$] in the presence of wildtype Hsp104.

The next major breakthrough in our understanding of [*PSI*$^{+}$] was the realization that assembly of Sup35 and certain fragments *in vitro* can provide a model for the replication of the prion state of Sup35 [24,25]. Purified full-length Sup35 protein, or a fragment containing the N-terminal 254 amino acids (NM region) will assemble into fibres after a lag phase in which the protein remains soluble [24]. These fibres share many characteristics with the

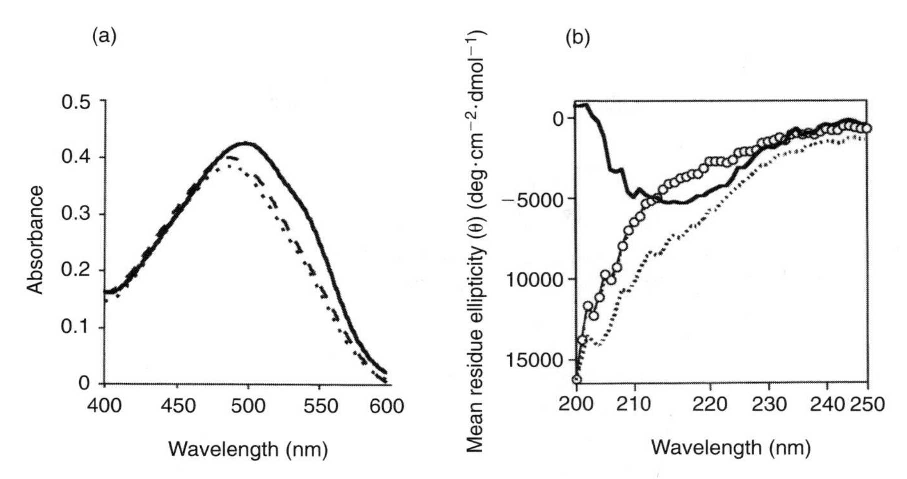

Figure 2 The NM fragment of Sup35p adopts an alternative conformation when assembled into amyloid fibres. (a) Congo Red binding. NM (1 μM) was added to a solution of Congo Red (10 μM) at room temperature. The absorption spectra of Congo Red alone (•••) or in the presence of NM fibres (—) or unpolymerized NM (- - -) are shown. (b) Circular dichroism. The circular dichroism spectra of NM alone, immediately after dilution from denaturant into aqueous solution (•••) or after 150 min (○). The addition of pre-assembled NM fibres (0.1 μM) to the buffer directs the conformational conversion of freshly diluted NM within 150 min (—).

protein amyloids associated with the TSEs, and other human diseases, including the capacity to bind to the amyloid diagnostic dye Congo Red and induce a shift in its absorbance spectrum (Figure 2A). In addition, Sup35 fibres are rich in β-sheet structure (Figure 2B), as is the case for other amyloids [26].

The key feature of fibre formation that links it to [*PSI*$^+$] biology is the ability of small quantities of preformed fibres to accelerate greatly the polymerization of soluble protein (Figure 2B). The conformational information present in Sup35 fibres is replicated *in vitro* using soluble Sup35 as a substrate.

Several observations link the process of amyloid formation *in vitro* to the propagation of [*PSI*$^+$]. First, lysates from [*PSI*$^+$] strains, but not [*psi*$^-$] strains, accelerate the conformational conversion and assembly of soluble NM region *in vitro* [24]. Secondly, the region of the protein that is crucial for fibre formation is also necessary for the inheritance of [*PSI*$^+$]. The Sup35 sequence has been divided into three regions, N (amino acids 1–125), M (amino acids 126–254), and C (amino acids 255–685), based on composition and similarity to other proteins [27–29]. Only fragments of Sup35 containing the N-terminal 125 amino acids (N) will form amyloid *in vitro* [24,25], and this region is both necessary and sufficient for the propagation of [*PSI*$^+$] *in vivo* [30]. Thirdly, like full-length Sup35, a fragment containing the N and M regions can exist in different states (soluble or insoluble) in [*psi*$^-$] and [*PSI*$^+$] strains (Figure 1) [19, 20]. *In vitro*, NM forms amyloid fibres, but can also exist in a soluble form for days [24]. Finally, mutations that alter the [*PSI*] status alter fibre formation in a

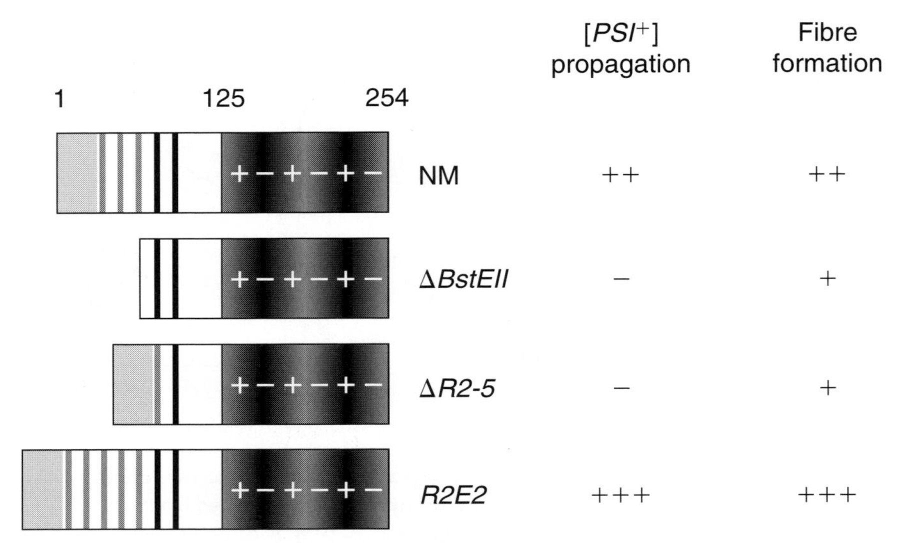

Figure 3 Mutations in the N region of Sup35 affect the propagation of [PSI^+] *in vivo* and the rate of fibre formation *in vitro*. Amino acids 1–41, grey shading; nonapeptide repeats 2–3, grey vertical bars; nonapeptide repeats 4–5, black vertical bars; amino acids 95–125, white box; amino acids 125–254, grey shading with + and −; ++, wild-type spontaneous appearance of [PSI^+] *in vivo* and rate of fibre formation *in vitro*; −, undetectable spontaneous rate of conversion from [psi^-] to [PSI^+] and inability to support [PSI^+] propagation; +, decreased rate of fibre formation *in vitro*; +++, accelerated rate of spontaneous [PSI^+] appearance *in vivo* and fibre formation *in vitro*. For additional information on the influence of mutations on fibre formation see references [24,31–33].

complementary fashion [14,24,31–33]. To provide just one example, the N region contains five imperfect repeats of the nonapeptide Gln-Gly-Gly-Tyr-Gln (Gln) Gln-Tyr-Asn-Pro [27–29]. Deletion mutants that have fewer copies of these repeats (Δ*BstEII* and Δ*R2-5*; see Figure 3) are unable to support [PSI^+] inheritance *in vivo* and form amyloid fibres at a reduced rate *in vitro* [24,31]. A mutant that has extra copies of the second nonapeptide repeat (*R2E2*; see Figure 3) has an increased rate of spontaneous appearance of [PSI^+] *in vivo*, and forms amyloid fibres at an accelerated rate *in vitro* [31]. Taken together, these observations indicate that the formation of amyloid fibres by Sup35 accurately recapitulates the propagation of [PSI^+] *in vivo*.

Nucleated conformational conversion

We have utilized the process of amyloid fibre formation *in vitro* to study the mechanism by which proteins replicate their conformational information. Three models have previously been proposed to explain this process [16,17,34,35]. The first model predicts that conformational information is

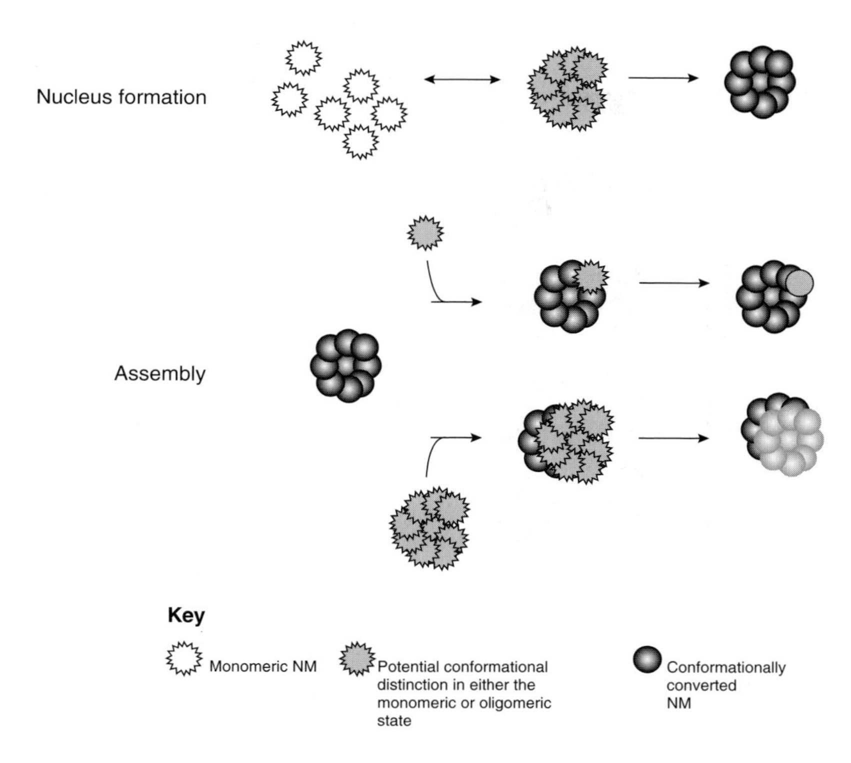

Figure 4 Nucleated conformational conversion model for NM fibre formation *in vitro*. Details of the model are presented in the text.

transmitted between monomeric isoforms [17]. The other two models predict that larger protein complexes are the active species; however, one model suggests that complexes catalyse polymerization of only one type of conformer [35], and the other model predicts that complexes promote conformation conversion during assembly [34].

Using a series of biochemical and microscopic criteria, we have analysed the process of amyloid formation by the NM fragment of Sup35 *in vitro* and propose a model for replication of this conformation information: nucleated conformation conversion (see Figure 4) [36]. In solutions of pre-formed NM fibres, the species that promotes (e.g. nucleates) the assembly and conformational change of soluble protein is sedimentable at 100 000 *g*, suggesting that solid-state NM contains the active conformational information. The efficiency of this nucleation is enhanced if fibres are fractured by sonication, suggesting that conformational information is transmitted through fibre ends.

Several lines of evidence indicate that conformational conversion is catalysed during assembly when soluble protein interacts with structured fibre ends.

(i) Conformational change cannot be temporally distinguished from assembly by a series of biochemical probes, including Congo Red binding, limited proteolysis, 8-anilino-1-naphthalene sulphonic acid binding, and circular dichroism in reactions proceeding at vastly different rates (minutes, hours, days).

(ii) The rate of assembly reaches a limit at higher monomer concentrations. We believe that this limit is imposed by the time required for protein interacting with the ends of fibres to conformationally convert to regenerate a surface that is competent for assembly. If fibres acted as nuclei for polymerization alone, assembly should be linearly dependent on soluble protein concentration.

(iii) Assembly time also reaches a limit if different methods that accelerate the process (e.g. fracturing the seed or increasing the collision frequency by agitation) are combined. Again, if fibres catalysed polymerization only, these treatments would be expected to result in an additive effect.

Formation of nuclei *de novo* and assembly onto pre-formed fibres is likely to proceed via an unstructured oligomeric intermediate.

(i) Oligomers of NM are observed by both transmission electron microscopy and atomic force microscopy (AFM). Mature fibres appear segmented by AFM, and the size of oligomers corresponds to that of the segments. In addition, oligomer–oligomer and oligomer–fibre interactions are observed frequently.

(ii) Conditions that alter the assembly of NM immediately after dilution from denaturant (e.g. solution, temperature or modifications) have a complementary effect on the rate of fibre formation.

(iii) The rate of assembly becomes limiting at concentrations where monomers are in excess over polymerization surfaces. However, the concentration of oligomers is likely to be limiting at this point.

These observations provide a model for the mechanism by which conformational information is transmitted to new proteins. The identification of an intermediate that is oligomeric and conformationally distinct from the amyloid state has broad implications for the inheritance of [*PSI*$^+$] *in vivo*. Indeed, formation of such an intermediate provides another potential point of action for regulators of [*PSI*] metabolism, such as Hsp104, as well as for the nature of the [*PSI*$^+$] state *in vivo*. Smaller complexes of Sup35 may mediate the nonsense suppressor phenotype, and would be expected to be transmitted cytoplasmically to progeny during cell division more efficiently than larger complexes. If [*PSI*$^+$] improves the fitness of strains [37,38], it may have evolved to maximize transmission, as distinct from the formation of amyloid fibres by disease-associated proteins in mammals. The *in vitro* system described here provides a method of assessing the mechanisms by which both intragenic and extragenic modifiers of [*PSI*$^+$] influence the process of conformational replication, and the delicate balance between states that allows these remarkable proteins to act as elements of inheritance.

References

1. Cox, B. (1965) Heredity **20**, 505–521
2. Tuite, M.F., Mundy, C.R. and Cox, B.S. (1981) Genetics **98**, 691–711
3. Lund, P.M. and Cox, B.S. (1981) Genet. Res. **37**, 173–182

4. Fink, G.R. and Conde, J. (1976) in International Cell Biology 1976–77: Papers Presented at the First International Congress on Cell Biology (Brinkley, B.R. and Porter, K.R., eds), pp. 414–419, Rockefeller University Press, New York
5. Cox, B.S., Tuite, M.F. and Mundy, C.J. (1980) Genetics **95**, 589-609
6. Cox, B.S., Tuite, M.F. and McLaughlin, C.S. (1988) Yeast **4**, 159–178
7. Serio, T.R. and Lindquist, S.L. (1999) Annu. Rev. Cell Dev. Biol. **15**, 661–703
8. Hawthorne, D.C. and Leupold, U. (1974) Curr. Top. Microbiol. Immunol. **64**, 1–47
9. Inge-Vechtomov, S. and Andriavnova, V. (1970) Genetika **6**, 103–115
10. Stansfield, I., Jones, K.M., Kushnirov, V.V., Dagkesamanskaya, A.R., Poznyakovski, A.I., Paushkin, S.V., Nierras, C.R., Cox, B.S., Ter-Avanesyan, M.D. and Tuite, M.F. (1995) EMBO J. **14**, 4365–4373
11. Zhouravleva, G., Frolova, L., Le Goff, X., Le Guellec, R., Inge-Vechtomov, S., Kisslev, L. and Philippe, M. (1995) EMBO J. **14**, 4065–4072
12. Chernoff, Y.O., Inge-Vechtomov, S.G., Derkatch, I.L., Ptysuhkina, M.V., Tarunina, O.V., Dagkesamanskaya, A.R. and Ter-Avanesyan, M.D. (1992) Yeast **8**, 489–499
13. Chernoff, Y.O., Derkach, I.L. and Inge-Vechtomov, S.G. (1993) Curr, Genet. **24**, 268–270
14. Derkatch, I.L., Chernoff, Y.O., Kushnirov, V.V., Inge-Vechtomov, S.G. and Liebman, S.W. (1996) Genetics **144**, 1375–1386
15. Wickner, R.B., (1994) Science **264**, 566–569
16. Griffith, J. (1967) Nature (London) **215**, 1043–1044
17. Prusiner, S.B. (1982) Science **216**, 136–144
18. Prusiner, S.B. (1998) Proc. Natl. Acad. Sci. U.S.A. **95**, 13363–13383.
19. Patino, M.M., Liu, J.J., Glover, J.R. and Lindquist, S.L. (1996) Science **273**, 622–626
20. Paushkin, S.V., Kushnirov, V.V., Smirnov, V.N. and Ter-Avanesyan, M.D. (1996) EMBO J. **15**, 3127–3134
21. Chernoff, Y.O., Lindquist, S.L., Ono, B., Inge-Vechtomov, S.G. and Liebman, S.W. (1995) Science **268**, 880–884
22. Parsell, D.A., Kowal, A.S., Singer, M.A. and Lindquist, S. (1994) Nature (London) **372**, 475–478
23. Glover, J.R. and Lindquist, S. (1998) Cell **94**, 73–82
24. Glover, J.R., Kowal, A.S., Schirmer, E.C., Patino, M.M., Liu, J.J. and Lindquist, S. (1997) Cell **89**, 811–819
25. King, C.Y., King, C.Y., Tittmann, P., Gross, H., Gebert, R., Aebi, M. and Wuthrich, K. (1997) Proc. Natl. Acad. Sci. U.S.A. **94**, 6618–6622
26. Serpell, L.C., Sunde, M. and Blake, C.C. (1997) Cell. Mol. Life Sci. **53**, 871–887
27. Kushnirov, V.V., Ter-Avanesyan, M.D., Telckov, M.V., Surguichov, A.P., Smirnov, V.N. and Inge-Vechtomov, S.G. (1988) Gene **66**, 45–54
28. Kikuchi, Y., Shimatake, H. and Kikuchi, A. (1988) EMBO J. **7**, 1175–1182
29. Wilson, P.G. and Culbertson, M.R. (1988) J. Mol. Biol. **199**, 559–573
30. Ter-Avanesyan, M.D., Dagkesamanskaya, A.R., Kushnirov, V.V. and Smirnov, V.N. (1994) Genetics **137**, 671–676
31. Liu, J.J. and Lindquist, S. (1999) Nature (London) **400**, 573–576
32. DePace, A.H., Santoso, A., Hillner, P. and Weissman, J.S. (1998) Cell **93**, 1241–1252
33. Kochneva-Pervakhova, N.V., Paushkin, S.V., Kushnirov, V.V., Cox, B.S., Tuite, M.F. and Ter-Avanesyan, M.D. (1998) EMBO J. **17**, 5805–5810
34. Uratani, Y., Asakura, S. and Imahori, K. (1972) J. Mol. Biol. **67**, 85–98
35. Jarrett, J.T. and Lansbury, Jr, P.T. (1993) Cell **73**, 1055–1058
36. Serio, T.R., Cashikar, A.G., Kowal, A.S., Sawicki, G.J., Moslehi, J.J., Serpell, L., Arnsdorf, M.F. and Lindquist, S.L. (2000) Science **289**, 1317–1321
37. Eaglestone, S.S., Cox, B.S. and Tuite, M.F. (1999) EMBO J. **18**, 1974–1981
38. Lindquist, S. (1997) Cell **89**, 495–498

Biochem. Soc. Symp. **68**, 45–68
(Printed in Great Britain)

4

Contribution of molecular chaperones to protein folding in the cytoplasm of prokaryotic and eukaryotic cells

Dean J. Naylor and F.-Ulrich Hartl[1]

Department of Cellular Biochemistry, Max-Planck-Institut für Biochemie, Am Klopferspitz 18A, Martinsried bei München D-82152, Germany

Abstract

While it is clear that many unfolded proteins can attain their native state spontaneously *in vitro*, the efficiency of such folding is usually limited to conditions far removed from those encountered within cells. Two properties of the cellular environment are expected to enhance strongly the propensity of incompletely folded polypeptides to misfold and aggregate: the crowding effect caused by the high concentration of macromolecules, and the close proximity of nascent polypeptide chains emerging from polyribosomes. However, in the living cell, non-productive protein folding is in many, if not most, cases prevented by the action of a highly conserved set of proteins termed molecular chaperones. In the cytoplasm, the Hsp70 (heat-shock protein of 70 kDa) and chaperonin families of molecular chaperones appear to be the major contributors to efficient protein folding during both normal conditions and adverse conditions such as heat stress. Hsp70 chaperones recognize and shield short, hydrophobic peptide segments in the context of non-native polypeptides and probably promote folding by decreasing the concentration of aggregation-prone intermediates. In contrast, the chaperonins interact with and globally enclose collapsed folding intermediates in a central cavity where efficient folding can proceed in a protected environment. For a number of proteins, folding requires the co-ordinated action of both of these molecular chaperones.

[1]To whom correspondence should be addressed.

Introduction

How do proteins fold in the cell?

The process by which a linear polypeptide attains its unique, functionally active, three-dimensional structure from the astronomical number of possible conformations has perplexed researchers for many years [1]. Most of our knowledge concerning the folding and assembly of proteins has come from *in vitro* studies with purified globular proteins under well-defined conditions. In their seminal studies, Anfinsen and colleagues observed that ribonuclease A, which had been unfolded by chemical denaturation, could refold spontaneously into its correct functional conformation upon dilution of the denaturing agent [2]. These experiments established that the information that specifies the native state of a protein resides in the amino acid sequence of the linear polypeptide chain — the 'principle of self-assembly'.

Although Anfinsen's principle of self-assembly remains unquestioned, researchers have recognized more recently that the success of protein folding *in vitro* is usually restricted to small, single-domain proteins, and often requires the use of low protein concentrations and low temperatures to decrease the probability that inappropriate interactions between exposed hydrophobic surfaces lead to aggregation. In contrast, protein folding and assembly *in vivo* is usually highly efficient, with greater than 95% of *de novo* synthesized polypeptides reaching their functional states [3,4]. This high efficiency is reached despite the requirement of many cell types to grow over a range of high temperatures and to maintain very high concentrations of proteins and other macromolecules, e.g. ~340 g/l in the *Escherichia coli* cytosol [5]. The high concentration of macromolecules within a cell occupies a large fraction of the intracellular space, resulting in 'macromolecular crowding'. This condition of crowding is believed to cause an increase in the intermolecular association constants of unfolded proteins that would increase significantly their tendency to aggregate, and thus limit the efficiency of their folding [6]. Recently this prediction has been confirmed by performing protein refolding experiments *in vitro* in the presence of polymers that mimic the macromolecular crowding effects of the cytoplasm [7]. Furthermore, the organization of translating ribosomes into polysomes should also impede efficient protein folding, as the close proximity of nascent chains would, again, increase their tendency to form unfavourable inter-molecular interactions. It is important to note in this context that nascent, i.e. ribosome-bound, polypeptide chains are thought to be topologically restricted by the ribosome and, thus, to retain aggregation-sensitive, unfolded structures; they cannot adopt stable tertiary folds until at least a complete, co-operatively folding polypeptide domain (~100–300 amino acids) has been synthesized and has emerged from the ribosome [8,9]. This effect is due to the fact that the C-terminal (~30 amino acids) residues of a nascent chain are within the ribosomal exit tunnel and are thus unavailable for folding [8,9]. Once completed and extruded from the ribosome, folding of a domain can proceed co-translationally in principle [10,11], thus avoiding possible incorrect intra-molecular interactions between concomitantly folding domains [10]. However, the efficiency of co-translational and sequential domain folding has been noted to be greater in eukaryotes than in bacteria, at least for certain types of proteins [10]. How do cells avoid a scenario where the majority of newly synthesized polypeptides aggregate? What then prevents these *de novo* synthesized polypeptides from associating inappropriately with themselves or other macromolecules during their folding? Numerous biochemical and genetic studies indicate that a pre-existing protein machinery, the 'molecular chaperones', fulfil such a role.

The molecular chaperone concept

As will be outlined in detail later, molecular chaperones essentially prevent protein misfolding and aggregation by binding to and stabilizing non-native protein conformations, before releasing them in a reaction often requiring ATP hydrolysis [12–14]. Chaperones acting in *de novo* folding, in general, recognize and shield exposed hydrophobic side chains, which are usually buried within the protein's native conformation, but otherwise have a tendency to associate inappropriately with other hydrophobic surfaces [12–14]. Therefore, the binding of chaperones to interactive surfaces in unfolded, but usually not native, proteins not only protects unfolded protein conformers from aggregation, but most importantly, through cycles of timed release and rebinding, also allows productive interactions to occur which are required for correct folding. In the event that such productive interactions do not allow the native state to be reached, the same or a different type of chaperone can recognize and (re)bind the unfolded polypeptide, permitting another opportunity for productive interactions to occur [12–14]. In this manner, different types of chaperones can work sequentially with synergistic effects. Through this process of binding and release, molecular chaperones typically increase the yield rather than the rate of protein folding. It should be noted that chaperone action is not solely confined to 'quality control' of protein folding, but also includes other essential cellular processes, such as intracellular sorting and membrane translocation of proteins [12–14].

The discovery of molecular chaperones by no means contradicts Anfinsen's principle of self-assembly [2]. Molecular chaperones do not convey steric information specifying the correct folding pathway of a non-native protein and do not form a structural component of the folded and assembled state; instead they allow efficient 'assisted' self-assembly to proceed in the unfavourable folding environment of a cell [15]. Furthermore, molecular chaperones do not require metabolic energy to actively promote protein folding, as the energy released through ATP hydrolysis by some chaperones permits the tightly regulated and timely release of the chaperone from its substrate. This is entirely consistent with Anfinsen's finding that protein folding is an energetically favourable reaction.

Using the nomenclature first described for *Drosophila*, most molecular chaperones are referred to on the basis of their ability to be induced by heat shock, and their apparent subunit molecular mass (kDa), as revealed by SDS/PAGE analysis [13,16]. Thus, the major families of heat shock proteins (Hsps) or molecular chaperones are: Hsp110, Hsp100, Hsp90, Hsp70, Hsp60 (chaperonins) and sHsp (small Hsp, with molecular masses between 15 and 30 kDa)

(Table 1). The amino acid sequence identity is high between members of each family but the identity between members within different families is usually insignificant. It should be noted that, although the problem of protein misfolding and subsequent aggregation increases with temperature, not all Hsps function as molecular chaperones. Some Hsps are proteases which co-operate with chaperones to remove unfolded or damaged proteins from the cell. Conversely, not all molecular chaperones are heat-stress inducible. Many molecular chaperones are, in fact, constitutively expressed to perform essential housekeeping functions at all temperatures encountered. One such function is, obviously, *de novo* protein folding. The constitutively expressed Hsps are often given the prefix Hsc, but this is not a strict rule. The focus of this review will be the mechanism of action by which the Hsp70 and chaperonin systems assist the folding of newly-synthesized polypeptides in the cytoplasm of prokaryotic and eukaryotic cells.

The Hsp70 molecular chaperone system

Conservation and diversification

The DnaK chaperone system of *E. coli* is regarded as the prototype Hsp70 system, and is composed of the DnaK molecular chaperone (a Hsp70 member) and the co-chaperones DnaJ (a Hsp40 member) and GrpE (~23 kDa). A multitude of studies have shown that homologues, and often multiple isoforms, of both DnaK and DnaJ exist in a large variety of prokaryotic cells and in multiple compartments of eukaryotic cells. Indeed, DnaJ homologues even exist in several tumour viruses [17–19]. The notable exception is the apparent absence of DnaK and DnaJ homologues from some, but not all, archaebacteria [20]. The conservation of the DnaK/DnaJ system is believed to have persisted throughout evolution because the co-operation with a DnaJ homologue is obligatory for the functioning of Hsp70 chaperones [18,19,21]. Numerous studies have shown that DnaJ-like proteins can target substrates to Hsp70 or recruit Hsp70 proteins to substrates. Thus, by increasing the number of DnaJ-like proteins, additional functional diversity of the Hsp70 system is generated [18,19]. This is probably the reason that *Saccharomyces cerevisiae* contains at least 17 DnaJ-like proteins compared with only 14 Hsp70 members [17]. In contrast, while GrpE homologues have been found in a large variety of prokaryotic cells and in one archaebacterium, they have only been detected in the mitochondria and possibly chloroplasts of eukaryotic cells [22–24] (Table 1). The absence of GrpE homologues from some Hsp70 systems might be explained by the action of additional co-factors that enable their Hsp70 partners to fulfil a broader range of tasks (discussed below). When considered together, these observations indicate that, during the course of evolution, the amplification of components for Hsp70 systems has been accompanied by a diversification of individual chaperone functions, where different Hsp70 teams could work either synergistically or independently to perform essential roles both during normal growth and in conditions of cellular stress [25].

Table 1 An overview of the distribution and subcellular location of molecular chaperones, co-chaperones and protein folding catalysts

Family	Prokaryotes (*E. coli*)	Eukaryotes		
		Cellular location	Fungi	Mammals
Hsp110	?	?	Sse1p (Msi3p), Sse2p	Apg-1 (Osp94), Apg-2 (Hsp70 RY)
(similar to Hsp70)		Nucleus	?	Hsp110 (Hsp105,107,112)
		Cytosol	Hsp88 (+mito. memb. bound)	Hsp110 (Hsp105,107,112)
		ER	?	Grp170
Hsp100 (Clps)	Class I:	?		ClpM-like (SKD3)
	Clp A,B,C,D,L	Cytosol	Hsp104 (ClpB homologue)	?
	Class II:	Nucleus	Hsp104 (ClpB homologue)	?
	Clp M,N,X,Y	Mito.	Hsp78 (ClpB homologue)	ClpX
			Mcx1p (ClpX homologue)	
Hsp90	HtpG (C62.5)	Cytosol	Hsp82, Hsc82	Hsp86 (Hsp90α), Hsp84 (Hsp90β)
		Nucleus	Hsp82	Hsp84 (Hsp90β)
		ER	?	Grp94 (ERp90, gp96, endoplasmin)
		Mito.	?	TRAP1 (Hsp75)
Hsp70	DnaK,	Cytosol	Ssa1-4p, Ssb1,2p,	Hsp72 (Hsp70), Hsp73
	Hsc66 (HscA),			(Hsc70, Prp73)
	and Hsc62	Nucleus	?	Hsp72, Hsp73
		ER	Ssd1p (Kar2p), Ssi1p (Lhs1p)	BiP (Grp78)
		ER memb.	?	Stch
		Mito.	mt-Hsp70 (Ssc1p),	mt-Hsp70 (Grp75, Pbp74)
			Ssh1p (Ssq1p),	
			Ssc2p (Ssj1p)	
		Mito. OM PA	?	Hsp73

Table 1 contd.

Family	Prokaryotes (*E. coli*)	Eukaryotes		
		Cellular location	Fungi	Mammals
Group I chaperonins	Eubacteria GroEL	Mito.	Hsp60 (Mif4p, Cpn60)	Cpn60 (Hsp60)
Group I co-chaperonins	Eubacteria GroES	Extracellular	?	EPF (Cpn10, Hsp10)
		Mito.	Cpn10 (Hsp10)	Cpn10 (Hsp10)
Group II chaperonins	Archaebacteria Thermasome (TF55)	Cytosol	TRiC (CCT)	TRiC (CCT)
Group II co-chaperonins	Archaebacteria GimC	Cytosol	GimC (Prefoldin)	Prefoldin (GimC)
Small Hsp (sHsp) 15–30 kDa	IbpA (14 kDa)	Mito. OM PA	Hsp30	?
	IbpB (16 kDa)	Cytosol	Hsp12, Hsp26, Hsp30	Hsp24 (25,27,28) + α-crystallin
		Nucleus	Hsp26	Hsp24 (25,27,28) + α-crystallin
Hsp40 (DnaJ)	DnaJ, CbpA, Hsc20 (HscB), Dj1A (RcsG)	?	Xdj1 (possibly a pseudogene), HLJ1	
		Nucleus	Zuotin, Sis1p	Hdj1 (Hsp40), Hdj2 (HSDJ), Hsj1a/b (neurone specific)
		Cytosol	Ydj1p (Mas5p), Sis1p, Zuotin, Djp1p, Caj1, (+memb. bound)	Hdj1 (Hsp40), Hdj2 (HSDJ), Auxillin Hsj1a/b (neurone specific), cysteine string protein, ALA-D
		ER	Scjlp	?

Table 1 contd.

Family	Prokaryotes (*E. coli*)	Eukaryotes		
		Cellular location	Fungi	Mammals
		ER memb.	Sec63p (Npl1p), Jem1p	? Mtj1 (Sec63p-like)
		Mito.	Mdj1p, Jaclp	hTid-1S, hTid-1L
		Mito. IM	Tim44, Mdj2p	mTim44
GrpE	GrpE	Mito.	Mge1p (Yge1, GrpEp)	mt-GrpE#1 + mt-GrpE#2
Immunophilins (PPlases)	10.1 kDa, PPI	Nucleus	?	CyP-40, FKBP25, FKBP52 (?)
	CyP18,21,	Cytosol	CYP1,2, FKB1 (RBP1)	CyP-40, CyPA, FKBP12,52
	WHP,	ER	CYP2	CyPB, CyPC, FKBP13
	FKBP33	Mito.	CYP3 (CPR3, cyclophilin 20)	CyPD
		Cell surface	?	NK-TR
NAC and TF	TF	Cytosol	αNAC and βNAC	αNAC and βNAC
PBF, MSF, Mft52 and SecB	SecB	Cytosol	Mft52, MSF (BMH1 + BMH2)	PBF + MSF

The members of each family have been grouped according to their sequence and functional similarities, except for the NAC/TF and the PBF/MSF/SecB groups of proteins, which may fulfil similar roles in prokaryotes and eukaryotes despite the lack of sequence similarity. Alternative abbreviations for otherwise identical components are shown in parentheses. Extra information is also shown in parentheses. Key abbreviations are: TCP-1, T-complex polypeptide 1; TRiC, TCP-1 ring complex; CCT, chaperonin containing TCP-1; PPlases, peptidylol-prolyl *cis/trans* isomerase; NAC, nascent polypeptide associated complex; mito., mitochondria; ER, endoplasmic reticulum; memb., membrane; OM, outer membrane; IM, inner membrane; PA, peripheral associated; TF, trigger factor; PBF, presequence binding factor; MSF, mitochondrial import stimulation factor. All of the references from which these data were derived can be found in [13].

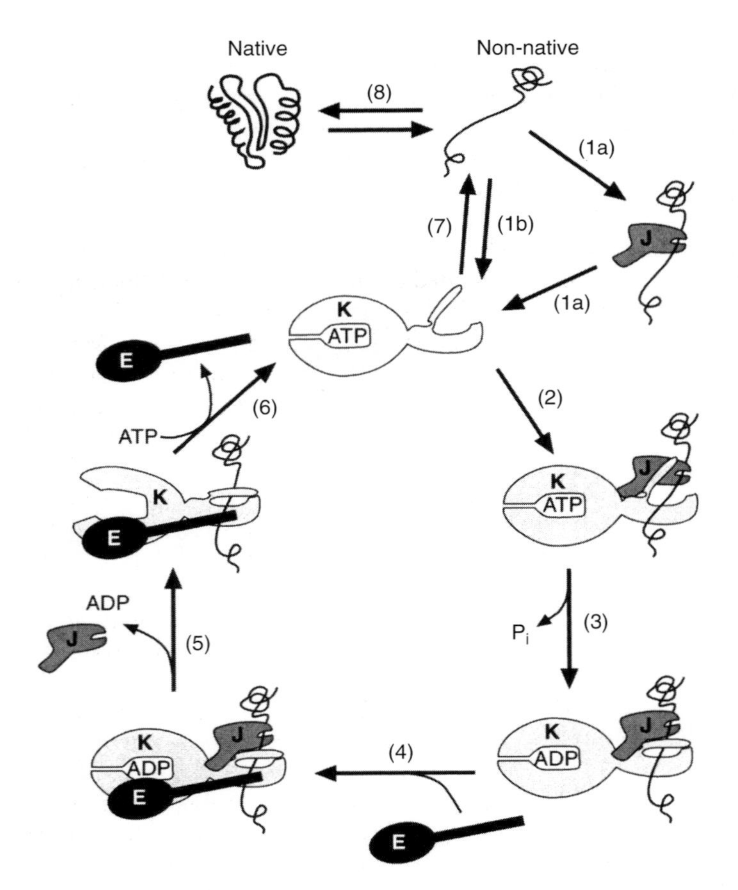

Figure 1 Proposed mechanism by which DnaJ (J) and GrpE (E) regulate the function of DnaK (K). The cycle begins with the association of a non-native protein with either DnaJ (1a) or ATP-bound DnaK (1b). Regardless of which chaperone binds the substrate initially, (2) the interaction of both the substrate and DnaJ with ATP-bound DnaK is required for efficient stimulation of the DnaK ATPase activity. (3) ATP hydrolysis is accompanied by a large conformational change in DnaK, which enables the substrate to be bound more tightly. (4, 5) GrpE interacts with the stable ADP-bound DnaK–substrate–DnaJ complex and induces the release of ADP by wedging apart DnaK's nucleotide-binding pocket. DnaJ leaves the complex as a result of the action of GrpE. (6) The rapid binding of ATP to DnaK is accompanied by another large conformational change which permits GrpE to leave the complex and reduces DnaK's affinity for substrate. (7) The substrate is released and either folds (8) or re-enters the cycle by associating with either DnaJ (1a) or DnaK (1b).

Structure and substrate specificity

Hsp70 molecular chaperones consist of two major domains: a 44 kDa N-terminal ATPase domain and a 25 kDa C-terminal substrate-binding domain, which communicate to mediate efficient binding and timely release of substrates (Figure 1). X-ray crystallographic structures of both isolated domains have been obtained; however, the structure of the intact Hsp70 molecule has yet to be determined. Co-crystallization and structural determination of the DnaK substrate-binding domain, in association with the peptide Asn-Arg-Leu-Leu-Leu-Thr-Gly, revealed that the C-terminal domain is composed of a further β-sandwich subdomain which houses a substrate-binding cavity, followed by an α-helical subdomain that acts as a lid to encapsulate the peptide within the cavity [26]. Early studies established that DnaK and eukaryotic Hsp70 members preferentially bind heptameric peptides containing large hydrophobic residues in an extended conformation [27–31]. Such segments are typically exposed only by proteins in a non-native state, explaining the ability of Hsp70s to distinguish between folded and unfolded proteins. More recently, a systematic peptide scanning approach was applied to define more rigorously the substrate specificity of DnaK [32]. A 'high-affinity' motif is composed of a 4–5 residue hydrophobic core enriched in Leu, but also in Ile, Val, Tyr and Phe, and flanked by two basic segments enriched in Arg and Lys. Acidic residues (Asp and Glu) are disfavoured throughout the motif [26,32]. The basic residues of the motif are thought to form electrostatic interactions with the negatively charged surroundings of the DnaK substrate-binding channel. The DnaK binding motif is statistically predicted to occur every 33 residues within protein sequences, which is consistent with the promiscuity with which DnaK (and other Hsp70 chaperones) binds to non-native protein substrates [32,33].

The residues contributing to the interaction of the peptide Asn-Arg-Leu-Leu-Leu-Thr-Gly with DnaK and those forming the β-subdomain in general (which houses the substrate-binding channel) are highly conserved amongst Hsp70 members [26,33]. In comparison, the residues contributing to the negative surface charges on either side of DnaK's substrate-binding channel are less conserved, and accordingly, differences in these surface charges have been predicted to influence substrate interactions with several Hsp70 members [26]. It is, therefore, not surprising that, although some divergent peptide-binding specificities have been observed amongst members of the Hsp70 family, the similarities are much more pronounced [30,34]. Thus, the described DnaK binding motif is probably similar for most Hsp70 members, although some differences are expected.

Reaction cycle of the DnaK (Hsp70)/DnaJ (Hsp40)/GrpE system

A combination of detailed biochemical, genetic and structural studies have revealed that the chaperone activity of Hsp70 proteins is manifested through cycles of substrate binding and release. Furthermore, such cycling requires large conformational changes in Hsp70 that are driven by ATP binding and hydrolysis and are tightly controlled by substoichiometric amounts of co-factors (or 'co-chaperones') [14,33]. The process is often termed the Hsp70 'ATPase cycle', and in its simplest form the cycle is described as an alternation between two conformational states: the ATP-bound state, with low affinity and high association/dissociation rates for substrates (where the substrate-binding channel is open), and the ADP-bound state, with high affinity and low association/dissociation rates for substrates (where the substrate-binding channel is closed) [14,33]. An understanding of how these two conformational states interconvert to allow Hsp70 proteins to interact transiently with substrates is best understood for the bacterial DnaK system (Figure 1). In this case, substoichiometric amounts of DnaJ and GrpE are required for the proper functioning of DnaK and are believed to regulate DnaK's ATPase cycle in a highly dynamic manner [35–37]. In fact, together DnaJ and GrpE can increase the ATPase activity of DnaK by at least 240-fold, which is probably more than sufficient to support its chaperone activity [38–40]. Analysis of the kinetic properties of the two conformational states of DnaK reveals that, in order to support efficient protein folding, DnaK must bind rapidly to unfolded proteins in its ATP-bound state, because, in the ADP-bound state, interaction would be too slow to prevent the aggregation of misfolded substrates. Furthermore, hydrolysis of ATP is then necessary to permit the transition of DnaK to the ADP-bound state, which is accompanied by a large conformational change and, subsequently, the stabilization of substrate interaction [39,41–47]. The intrinsic ATPase activity of DnaK ($k_{cat} \approx 0.04$ mol of ATP hydrolysed/mol of DnaK per min, at 30°C), however, is not high enough to allow efficient stabilization of the substrate–DnaK interaction before the substrate dissociates [39,48,49]. This problem is overcome by the synergistic stimulation of the ATPase activity of DnaK (and other Hsp70s) by DnaJ and a polypeptide substrate [21].

After a suitable interaction time of the substrate with DnaK, the substrate is released in a process that requires the exchange of bound ADP for ATP and is coupled to the opening of the peptide-binding cleft [14,33]. Although the release of ADP from DnaK has been shown to be 10–20-fold faster than the rate of ATP hydrolysis in an unstimulated ATPase cycle, it becomes rate-limiting in the DnaJ-stimulated cycle, because in the presence of substrate (but absence of GrpE) DnaJ appears to stabilize the ADP-bound state [37,39,48]. In *E. coli* this limitation is overcome with the assistance of GrpE, which accelerates ADP release by up to 5000-fold [50], allowing the rapid binding of ATP [48,51]. Unlike the original proposal [41], conformational changes resulting from ATP binding, rather than hydrolysis, trigger substrate dissociation, thereby completing the DnaK ATPase cycle and permitting DnaK to enter another round of substrate binding and release [46–48]. In summary, the function of DnaJ is to facilitate the association and stable binding of polypeptide substrate to DnaK, whereas GrpE facilitates substrate release. Together, DnaJ and GrpE can tightly regulate the duration of the substrate–DnaK interaction.

Although a transient ATP-dependent interaction between DnaJ and DnaK has been detected, such complexes are unstable and therefore cannot be isolated [52]. Other studies have isolated ternary DnaJ–substrate–DnaK complexes, where DnaK is, presumably, in the ADP-bound state, although it is unknown whether the association of DnaJ with such complexes is through its ability to bind unfolded polypeptides or its ability to interact with DnaK directly [47,53–57]. In contrast, GrpE binds strongly to DnaK, and while the interaction is persistent in 2 M KCl, the complex is rapidly dissociated in ATP [58]. In fact, the specificity of this interaction has been utilized to isolate mitochondrial GrpE-like molecules from a variety of organisms [59–62]. The X-ray crystallographic structure of an N-terminally truncated (residues 34–197) GrpE bound to the ATPase domain of DnaK (residues 3–383) has been recently solved [63]. This confirmed previous studies which showed that a dimer of GrpE binds to a single molecule of DnaK in the absence of ATP [64–66]. While the structural topology of each GrpE monomer is similar, the elongated dimer is asymmetrically bent towards DnaK. One proximal monomer of GrpE is closest to and forms most of the contacts with six distinct surface areas of DnaK [63]. Within the GrpE–DnaK complex, the distal monomer of GrpE appears to function mainly in stabilization of the proximal monomer; together they form an unusually long parallel α-helical structure that extends towards DnaK's substrate-binding domain and is postulated to play a role in substrate release [63]. The long N-terminal α-helices (~100 Å) are followed by a shorter four-helix bundle, to which each monomer of GrpE contributes two helices. The two monomers end in a six-stranded β-sheet and they extend from their respective four-helix bundle like a pair of arms. The proximal β-sheet arm is proposed to facilitate release of DnaK-bound ADP by wedging out the IIB subdomain of DnaK by 14 Å, relative to its position in the ADP-bound structure. The wedging action displaces, by as much as 3 Å, three DnaK residues (Glu-267, Lys-270 and Ser-274) that provide important hydrogen bonds to the adenine and ribose rings of bound ADP. This explains the absence of bound ADP in the crystal structure defining the interaction between GrpE and the ATPase domain of DnaK compared with the ADP-bound structure of the bovine Hsc70 ATPase domain [63].

In some Hsp70 systems, GrpE may be replaced by additional co-factors that broaden the functions of Hsp70 chaperones

Given that the Hsp70 (DnaK)/Hsp40 (DnaJ) system has been highly conserved throughout evolution, it is intriguing that GrpE homologues may not be required for the efficient functioning of this system in the cytosol, the nucleus or the endoplasmic reticulum of eukaryotic cells. In agreement with this contention, it is known that the rate-limiting step in the prokaryotic (DnaK/DnaJ) and mitochondrial Hsp70/Hsp40 systems is the release of bound ADP, a process facilitated by GrpE [38,47,51]; in contrast, the rate-limiting step for the cytosolic/nuclear yeast (Ssa1p/Ydj1p) and mammalian Hsp70/Hsp40 systems is ATP hydrolysis [67–69]. In the latter cases, ADP is released spontaneously, and therefore GrpE may be dispensable for these systems.

There are two problems with this scenario though. First, the ADP-bound state of cytosolic/nuclear Hsp70 is too short-lived to permit an efficient interaction with a substrate [68,69]. Secondly, recent studies have identified several co-factors that are capable of regulating the cytosolic/nuclear Hsp70 system although their precise roles are still largely unknown. For example, a Hsc70-interacting protein (Hip) (or p48) of approx. 41 kDa prevents the release of ADP from Hsc70 and, thereby, stabilizes substrate binding [68]. Another protein, BAG-1 (Hap, Hap46, RAP46), competes with Hip for binding to the Hsc70 ATPase domain, and, once bound, can stimulate the ATPase activity of Hsc70. The mechanism by which BAG-1 stimulates the ATPase activity of Hsc-70 is controversial; it accelerates either ADP release [70] or ATP hydrolysis [71]. As a result, BAG-1 has been observed to both stimulate [72] and prevent [71] the release of substrates from Hsc70. These differences may, at least partly, be attributed to the study of different BAG-1 isoforms. Recently, an evolutionarily conserved family of BAG-1 proteins has been identified, with at least five members in humans that range in molecular mass from approx. 36 to 58 kDa [73,74]. All members contain distinct N-terminal regions but share a conserved C-terminal BAG domain (of approx. 45 amino acids) which binds Hsp70 [74]. Like many Hsp40 members, BAG-1-like proteins have been found to interact with numerous polypeptides and may also serve as adaptors for recruiting Hsp70 members to perform chaperone functions in various cellular processes [74].

An additional Hsp70 adaptor is the approx. 60 kDa Hsc70/Hsp90-organizing protein (Hop), alternatively named p60, Sti1 and RF-Hsp70 [75]. Hop appears to target Hsp90 to complexes of Hsp70 with steroid receptors or oncogenic protein kinases, to facilitate the folding and functional maturation of these signal transduction proteins [75–77].

The chaperonin system

Conservation and diversification

The chaperonins constitute essential gene products that are found in all three domains of life and facilitate the ATP-dependent folding of *de novo* synthesized polypeptides and of stress-denatured proteins within the central cavities of their double-ring oligomeric structures. Based on their evolutionary lineage, the chaperonins are divided into two distantly related groups [78]: the group I chaperonins, found in plastids (Rubisco-binding protein or Cpn60), mitochondria (Hsp60 or Cpn60), and eubacteria (GroEL); and the group II chaperonins, found in archaebacteria (TF55 or Thermosome) and the cytosol of eukaryotic cells [TCP-1 (T-complex polypeptide 1) ring complex TRiC) or CCT] (Table 1). Group I chaperonins are usually homo-oligomeric structures with subunits of approx. 60 kDa that are composed of two heptameric rings arranged back-to-back. The exceptions are mammalian mitochondrial Hsp60 (Cpn60), which appears to be active as a single homo-heptameric ring, and plastid Cpn60, which contains two heptameric rings formed by two types of related subunits. They co-operate with smaller co-chaperones (GroES in *E. coli*, Hsp10 or Cpn10 in mitochondria) that form homo-heptameric rings from approx. 10 kDa subunits, with the exception of plastid Cpn21, whose subunits are composed of two Cpn10-like domains fused together in tandem [13,79]. These co-chaperones function as lids to enclose an unfolded protein of up to approx. 60 kDa within the central cavity of the group I chaperonins, where folding can proceed in a protected environment. Interestingly, bacteriophage T4 has its own co-chaperone (gp31) which can functionally substitute for GroES in *E. coli* [80]. By forming a specialized lid for GroEL, gp31 has been proposed to enlarge the GroEL central cavity to permit the efficient folding of the large-sized phage capsid protein (gp23) [80].

Group II chaperonins are hetero-oligomeric structures composed of distinct (but related) approx. 60 kDa subunits that also form a double ring arrangement stacked back-to-back. In the eukaryotic cytosol, TRiC is composed of eight distinct subunits per ring, while in the archaeal cytosol Thermosome rings have eight or nine subunits composed of one or two types.

Unlike group I chaperonins, group II chaperonins function without a GroES-like co-chaperone, probably because they contain their own 'built-in lid' in the form of helical protrusions [81,82]. Recently, a newly described hexameric chaperone GimC (or Prefoldin) has been reported to function in co-operation with TRiC and Thermosome [83–87]. In both yeast and mammals GimC is composed of six distinct (but related) subunits, while in archaebacteria GimC is only formed by two distinct (but related) subunits [83–85,87]. Unlike the GroES-like co-chaperones, GimC plays a more active role in protein folding by interacting with unfolded proteins and stabilizing them against aggregation for subsequent folding by TRiC or Thermosome. GimC has been reported to assist in the transfer of nascent polypeptide chains to TRiC *in vivo*, and may have a 'proofreading' function to prevent the premature release of non-native substrates from TRiC into the perilous cytosol [85,86].

Structure and function

The mechanism by which chaperonins assist in efficient protein folding is best understood for *E. coli* GroEL–GroES as the prototype group I chaperonin system. X-ray crystallographic structures of both the tetradecameric GroEL and the heptameric GroES alone have been attained [88–91]. The structures of GroEL, fully-occupied with 14 molecules of non-hydrolysable [γ-^{35}S]ATP, and most importantly, of the asymmetrical GroEL–GroES–(ADP)$_7$ complex have been determined also [92,93]. GroEL is composed of two heptameric rings of identical approx. 57 kDa subunits arranged back-to-back to form two central cavities that are open at the ends but closed between the interface of the rings (Figure 2). Each of the GroEL subunits consists of three distinct domains: an apical domain which forms the opening of the cavity and contains a hydrophobic substrate binding surface that faces the cavity of the GroES unoccupied ring; an equatorial domain that contains the ATP-binding site and forms all of the inter- and most of the intra-ring interactions; and an intermediate hinge-like domain that connects the apical and equatorial domains. The two crystal structures of the co-chaperone GroES reveal that it is a dome-shaped heptameric ring [90,91]. Each GroES subunit contains a mobile loop that extends from the base of the dome and becomes ordered upon interaction with a hydrophobic groove within each apical domain of GroEL [94]. The site of interaction has been shown in several studies to overlap partially with the site to which substrates are bound by GroEL [93,95–98]. Thus, GroES can bind as a 'lid' over the opening of the same ring of GroEL that contains the substrate (Figure 2).

GroES binding is preceded by the binding of seven ATP molecules to GroEL, which promotes the opening of the hydrophobic apical domains, by an upwards and outwards twisting motion, and the subsequent release of the substrate into the central cavity [93,96,99–105]. As a consequence of these ATP-induced apical domain movements, the volume of the so-called '*cis* cavity' (or 'Anfinsen cage') is approximately doubled (now approx. 80 Å in width and approx. 85 Å in length), and the hydrophobic binding sites are buried between GroEL subunits and within the interface with GroES [92,93,96]. As a result of these conformational changes, a hydrophilic lining of the cavity is created that

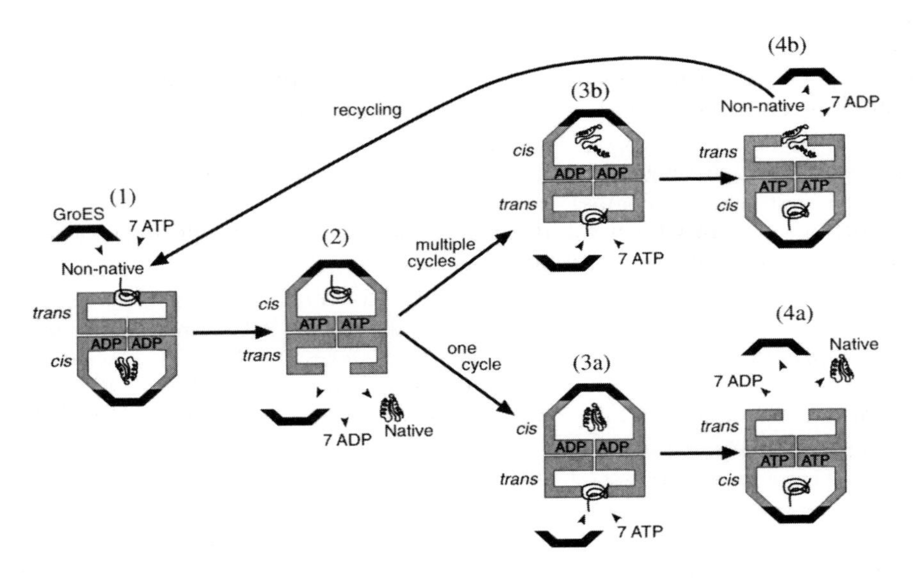

Figure 2 Schematic representation of the GroEL–GroES reaction cycle. GroEL–GroES-mediated protein folding involves the alternating action of both GroEL rings cycling in a co-ordinated manner. (1) The cycle begins with the association of an unfolded protein with the hydrophobic apical domains of the (upper) vacant *trans* ring of the asymmetrical GroEL–GroES complex. (2) Binding of ATP to the same (upper) ring then causes the hydrophobic apical domains to twist upwards and outwards, which effectively doubles the volume of the cavity and releases the bound substrate into the central cavity. Consequently, GroES is permitted to bind the now free (upper) hydrophobic apical domains and form a new *cis* complex where the captured substrate may fold in a protected environment. At the same time GroES is released from the lower ADP-bound *cis* cavity, thereby allowing the substrate to exit. (3a and 3b) In the newly formed *cis* complex (upper), ATP hydrolysis appears to function as a timer to permit productive folding to proceed for approx. 10–20 s before the substrate is released. (4a) The binding of substrate and ATP to the (lower) *trans* ring stimulates synergistically the release of GroES and the now native, folded substrate from the (upper) *cis* cavity, while (4b) substrates which require multiple cycles in order to attain their native states either may be rapidly rebound by the same GroEL molecule in preparation for the next folding cycle or may be released in a non-native state (not shown). The subsequent binding of GroES to the (lower) *trans* ring is concomitant with the release of the bound substrate into the central cavity and the formation a new (lower) *cis* ring, thus ending a single cycle while commencing a new one.

is conducive to the folding of an enclosed polypeptide substrate of up to approx. 60 kDa. Within the GroEL–GroES *cis* cavity [106,107], the substrate is then permitted to fold efficiently, according to the principle of self-assembly, in an environment free from the adverse inter-molecular interactions that prevail in the crowded cellular cytosol and cause aggregation [102,104]. However, besides preventing aggregation, GroEL may also have the ability to partially unfold a bound substrate by exerting an unspecific stretching force manifested through the ATP-induced apical domain movements which release the substrate prior to GroES binding [108]. It has been proposed that forceful unfolding is necessary to promote the folding of 'kinetically trapped' folding intermediates that, for unknown reasons, neither aggregate rapidly, nor fold spontaneously into their native states *in vitro*. Whether such kinetically trapped folding intermediates exist within cells, and whether forceful unfolding is actually a necessary feature of the GroEL mechanism for promoting productive folding, remains to be determined.

The lifetime of the substrate in the GroEL–GroES *cis* cavity is ultimately determined by the GroEL ATPase that serves as a timer. When viewed by electron microscopy, the asymmetrical complexes of GroEL–GroES have a bullet-like appearance (Figure 2) and, in the presence of either *cis* ATP or ADP, they are often referred to as ATP-bullets or ADP-bullets, respectively [96,105,109]. The ATP-bullet is considerably more stable than the corresponding ADP-bullet; thus ATP hydrolysis produces a labile ADP-bullet that is primed for the release of GroES and, subsequently, the substrate [100,110–112]. The release of GroES from the ADP-bullet is triggered by the binding of seven ATP molecules (not hydrolysis) to the unoccupied GroEL *trans* ring, and the ATP-dependent mechanism can be stimulated by at least 50-fold by the simultaneous binding of a second substrate to the *trans* ring [111–113]. Because the second substrate binds to the *trans* ring before GroES, the assembly of a new folding cavity would be coupled to the disassembly of the old one in the opposite ring. Thus, the GroEL–GroES system appears to function as a 'two-stroke motor', alternating its folding-active forms between rings [114].

A single GroEL–GroES folding/reaction cycle proceeds for approx. 10–20 s, i.e. the lifetime of the *cis* complex [110,112,115,116]. For many unfolded proteins this time is sufficient to attain their native state, but for others, only a fraction will complete folding in one cycle of the GroEL–GroES system. *In vitro* studies, concerning the fate of partially folded protein intermediates upon the release of the GroES lid (at the end of each cycle), have led to differing views about the mechanistic action of the GroEL–GroES system in the cell [12,117]. Several studies performed in dilute buffered solutions have observed that non-native intermediates can be released into the bulk solution [110, 118–122]. In the cell, such proteins are believed to fold spontaneously, to be rebound by another chaperone or, eventually, to be targeted for degradation. These studies have been interpreted in support of the so-called iterative annealing model for the action of GroEL–GroES, in which kinetically trapped folding-intermediates have to be unfolded in order to give them a new opportunity to fold correctly [117]. In the iterative annealing model, folding occurs either in the GroEL–GroES *cis* cavity or upon release into the bulk cytosol and the site of folding is not regarded as mechanistically significant. Those substrates that are discharged into the bulk cytosol in a state unable to fold productively can then be rebound by another GroEL molecule for a further round of unfolding. Another view, the Anfinsen cage model, argues that the release of unfolded proteins from GroEL into the bulk cytosol would favour their aggregation, and that such non-productive loss is prevented under the macromolecular crowding conditions of the cell [12,123]. Thus, the Anfinsen cage model of GroEL–GroES-assisted protein folding stipulates that folding-competent, non-native substrates will preferentially be re-bound by the same chaperonin molecule *in vivo*, and will only be released into the bulk cytosol once they have attained their native states or have reached a form capable of attaining their native states without significant danger of aggregation. The two models are not mutually exclusive, and it is possible that both mechanisms may operate on specific substrates *in vivo*. However, there is a strong consensus that, regardless of whether non-native protein is released from the chaperonin, the folding reaction proceeds in the GroEL–GroES cage [102–104].

Chaperonin substrate specificity

One of the most intriguing features of the chaperonin system is its ability to bind promiscuously to a large and diverse number of non-native proteins in a sequence-independent manner. Thus, approx. 50% of chemically denatured proteins from a soluble *E. coli* lysate will bind GroEL, and a significant fraction (approx. 30%) of *E. coli* proteins has been estimated to misfold in GroEL-deficient cells [124,125]. Approximately 300 out of a total of approx. 2500 cytosolic polypeptides interact transiently with GroEL upon synthesis [126,127]. Consistent with the essential function of GroEL in *E. coli* [128], identification of 52 of these GroEL substrates *in vivo* revealed that they constitute proteins involved in transcription/translation, DNA manipulation, cell division, a variety of metabolic processes and several additional important cellular functions [127]. The majority of the proteins that interact with GroEL *in vivo* were found to be smaller than 60 kDa (the capacity of the GroEL–GroES *cis* cavity) and a preference for proteins containing multiple αβ-domains was noted among the identified substrates [127]. Such domains contain extensive hydrophobic surfaces and are, therefore, expected to cause the proteins that contain them to be prone to aggregation [127].

The promiscuity with which GroEL binds non-native proteins has been attributed to the plasticity with which the hydrophobic binding sites of GroEL accommodate a variety of substrates [98]. Analysis of the crystal structures of a high affinity peptide, termed SBP (Ser-Trp-Met-Thr-Thr-Pro-Trp-Gly-Phe-Leu-His-Pro), in complex with an isolated apical domain of GroEL and with tetradecameric GroEL localized the site of interaction to a flexible groove within each GroEL apical domain [98]. In native GroEL these binding sites would appropriately line the openings of both central cavities where non-native substrates are bound. Furthermore, this constitutes the same binding site that is occupied by both the mobile loop of GroES (Glu-Thr-Lys-Ser-Ala-Gly-Gly-Ile-Val-Leu-Thr-Gly-Ser) in the crystal structure of the GroEL–GroES–$(ADP)_7$ complex, and a seven-residue N-terminal tag (Gly-Leu-Val-Pro-Arg-Gly-Ser) of a neighbouring molecule in the crystal structure of an isolated GroEL apical domain [93,129]. Comparison of the three bound peptides revealed that, while they adopt different conformations (β-hairpin for SBP, a loop for the mobile loop of GroES, and an extended structure for the N-terminal tag), an extended segment of each peptide provides most of the binding surface and adheres to the binding groove with the same N- to C-terminus polarity and hydrogen-bonding pattern [98]. Most significantly though, the conformation of the GroEL binding site is subtly different for each of the three bound peptides and also in every unliganded structure solved [98]. This suggests that, within the GroEL cavity, flexible binding sites can adjust individually to accommodate a diverse number of non-native proteins.

Interestingly, a recent cryo-electron microscopic reconstruction study revealed that specific subunits of the group II chaperonin TRiC are required for interaction with α-actin [130]. On the other hand, another report revealed that TRiC interacts with approx. 9–15% of newly synthesized cytoplasmic polypeptides ranging from 30 to 60 kDa in size [131]. Together, these studies suggest that while, TRiC probably binds with numerous non-native proteins, the divergence of the eight TRiC subunits may permit TRiC to interact with and fold a number of specific substrates such as actin. Indeed, although GroEL can bind actin rapidly, it cannot assist its productive folding [132].

Contribution of molecular chaperones to the folding of newly synthesized proteins and those unfolded by stress

Most of the molecular chaperone activity involved in *de novo* protein folding in the cytosol appears to be due to the Hsp70 and chaperonin systems [12,14,133]. For example, in the cytosol of mammalian cells, a minimum of approx. 15–20% and approx. 9–15% of newly synthesized cytoplasmic polypeptides were observed to interact transiently with Hsc70 and TRiC, respectively [131]. This would leave the possibility of a majority of *de novo* synthesized polypeptides reaching their native state without the assistance of the Hsp70 and chaperonin systems [131]. Although certain substrates, such as inactive steroid receptors and oncogenic protein kinases, require an interaction with the highly conserved Hsp90 chaperone for their folding in eukaryotic cells, Hsp90 does not appear to have a general role in *de novo* protein folding [133,134]. Furthermore, these Hsp90 substrates probably require an initial interaction with Hsc70. Some proteins, such as small single-domain polypeptides and larger proteins composed of small, independent domains, may fold co-translationally without the assistance of chaperones [9,10]. Other nascent polypeptides may be assisted in folding by the ribosomes themselves [135–138].

In comparison with the proportion of newly synthesized polypeptides observed to associate with Hsc70 and TRiC in the mammalian cytosol, a minimum of approx. 5–10% and approx. 10–15% of *de novo* synthesized polypeptides interact transiently with *E. coli* DnaK and GroEL respectively under normal growth conditions [126,139,140]. The proportion of newly synthesized cytoplasmic polypeptides associated with GroEL increases from approx. 10–15% to approx. 30% when *E. coli* cells are shifted from growth at 30°C to growth at 42°C [126]. In accordance with the overlapping functions of trigger factor (TF) and DnaK in *de novo* protein folding, in TF deletants the fraction of nascent polypeptides interacting with DnaK doubles [139]. Clearly, more work will be necessary to determine the quantitative contribution of chaperones to protein folding under a variety of conditions.

Co-operation of molecular chaperone systems in protein folding

Several studies have reported a successive and directional action of chaperone systems (i.e. Hsp70→chaperonin or Hsp70→Hsp90) to occur during the folding of certain newly synthesized polypeptides. These chaperones function synergistically by promoting early and later folding steps [75,133,141–148]. A recent study employing a *dnaK* deletion mutant revealed that although some proteins require the successive action of DnaK and GroEL for efficient folding *in vivo*, the initial interaction with DnaK is not obligatory for these proteins to interact with GroEL [139]. However, it appears that in the absence of DnaK, another chaperone, trigger factor, may be involved in the transfer of such polypeptide chains to GroEL [139]. In prokaryotic and eukaryotic cells, Hsp70 members interact co-translationally with a subset of nascent polypeptides as they emerge from ribosomes [139,146,147,149–153]. It is proposed that the initial interaction with Hsp70 maintains the folding competence of these translating polypeptides, which, in some cases, may commence folding co-translationally (i.e. for multi-domain proteins composed of independent modules). However, in other cases complete synthesis is required before folding can proceed (i.e. for single-domain proteins, multi-domain proteins composed of discontinuous amino acid sequences, and proteins that have to be post-translationally translocated across membranes). For a number of polypeptides the interaction with Hsp70 may be sufficient for folding and assembly, while for others the interaction is necessary but does not result in complete folding [139,143,145,154]. In the latter case, which often represents polypeptides that have difficulty reaching or maintaining their native conformations, these polypeptides require a further interaction with either a chaperonin or an Hsp90 system for the completion of their folding and assembly [9,134]. The sequential chaperone action is consistent with Hsp70 members interacting with earlier *de novo* folding-intermediates, whereas the chaperonins and Hsp90 members are proposed to recognize intermediates that have native-like secondary structure and, in some cases, a global topology that resembles the completed tertiary state [12,76,155]. While the successive and directional action of the Hsp70→Hsp90 systems has only been studied in the eukaryotic cytosol [75,133], the action of the Hsp70→chaperonin systems has been well documented in the mitochondrial matrix, *E. coli* cytosol and eukaryotic cytosol [139,141–148]. Nevertheless, both pathways are known to involve a number of co-chaperones for their efficient functioning [12,14,133].

In contrast to the view of a directional association of newly synthesized polypeptides with chaperone systems during *de novo* folding ('pathway model'), the so-called 'network-model' holds that folding *in vivo* is governed by free kinetic partitioning of non-native intermediates between the bulk cytosol and a 'chaperone network' [156–158]. For example, denatured luciferase, which cannot fold on GroEL, can be passed back and forth between the DnaK and GroEL systems *in vitro* [156]. However, it appears that there are differences between *de novo* protein folding *in vivo* and the refolding of denatured proteins *in vitro* [148]. Upon import into the mitochondrial matrix, firefly luciferase accumulated on the chaperonin Hsp60 in an unfolded state and did not utilize retrograde transfer to mitochondrial Hsp70 for its folding [148]. Furthermore, expression of a heterologous chaperonin-trap (GroEL D87K) that irreversibly binds non-native actin and other newly synthesized, but not yet folded, polypeptides in the cytosol of yeast and mammalian cells did not interfere with protein folding [85,131]. This indicates that *de novo* protein folding, at least for many proteins, may proceed without the release of unfolded states into the bulk cytosol. Specifically, the chaperonin trap could not interfere with folding of newly synthesized actin, which appears to require the successive action of Hsc70/Hsp40 and TRiC/GimC (alternatively named Prefoldin) [84,85,131,146,147]. It is proposed that this folding pathway is compartmentalized so that non-native actin is not released into the bulk cytosol and is prevented from forming unproductive interactions with macromolecules in the crowded cytosol [85,126,131]. In contrast, stress-denatured and folding-incompetent proteins are efficiently bound by the chaperonin trap and, therefore, are likely to require partitioning between chaperones and the bulk cytosol *in vivo* [85,131,159]. In addition, post-translational cycling between chaperones and the bulk cytosol may assist the assembly of oligomeric proteins [147].

Concluding remarks

While our understanding of the mechanism by which chaperones function to prevent protein aggregation and promote productive folding has advanced significantly in recent years based on studies *in vitro*, important questions concerning the role of chaperones in protein folding *in vivo* remain to be addressed. Following the recent completion of the human, *S. cerevisiae* and *E. coli* genome sequencing projects, it is now a challenge to identify the complete set of molecular chaperones within an organism and to understand how different chaperone systems contribute to and co-operate in the folding of that organism's proteome. Studies to define the set of *in vivo* substrates of specific chaperones are well under way and, given the rapid progress in this field, we will probably not have to wait long before some interesting results emerge. An understanding of the *in vivo* functions of molecular chaperones may also open up new therapeutic approaches to a number of neurodegenerative and other diseases that result from protein misfolding.

References

1. Dobson, C.M. and Karplus, M. (1999) Curr. Opin. Struct. Biol. **9**, 92–101
2. Anfinsen, C.B. (1973) Science **181**, 223–230
3. Gething, M.-J., McCammon, K. and Sambrook, J. (1986) Cell **46**, 939–950
4. Copeland, C.S., Doms, R.W., Bolzau, E.M., Webster, R.G. and Helenius, A. (1986) J. Cell Biol. **103**, 1179–1191
5. Zimmerman, S.B. and Trach, S.O. (1991) J. Mol. Biol. **222**, 599–620
6. Ellis, J. and Hartl, F.-U. (1996) FASEB J. **10**, 20–26
7. van den Berg, B., Ellis, R.J. and Dobson, C.M. (2000) EMBO J. **18**, 6927–6933
8. Hardesty, B., Tsalkova, T. and Kramer, G. (1999) Curr. Opin. Struct. Biol. **9**, 111–114
9. Netzer, W.J. and Hartl, F.-U. (1998) Trends Biochem. Sci. **23**, 68–73

10. Netzer, W.J. and Hartl, F.-U. (1997) Nature (London) **388**, 343–349
11. Nicola, A.V., Chen, W. and Helenius, A. (1999) Nat. Cell Biol. **1**, 341–345
12. Hartl, F.-U. (1996) Nature (London) **381**, 571–580
13. Gething, M.-J. (ed.) (1997) Guidebook to Molecular Chaperones and Protein-Folding Catalysts, Oxford University Press, New York
14. Bukau, B. and Horwich, A.L. (1998) Cell **92**, 351–366
15. Ellis, R.J. and Hartl, F.U. (2000) Curr. Opin. Struct. Biol. **10**, 13–15
16. Georgopoulos, C. and Welch, W.J. (1993) Annu. Rev. Cell Biol. **9**, 601–634
17. Laufen, T., Zuber, U., Buchberger, A., and Bukau, B. (1998) in Molecular Chaperones in the Life Cycle of Proteins (Fink, A.L. and Goto, Y. eds) pp. 241–274, Marcel Dekker, New York
18. Kelley, W.L. (1998) Trends Biol. Sci. **23**, 222–227
19. Kelley, W.L. (1999) Curr. Biol. **9**, 305-308
20. Gribaldo, S., Lumia, V., Creti, R., Conway de Macario, E., Sanangelantoni, A. and Cammarano, P. (1999) J. Bacteriol. **181**, 434–443
21. Laufen, T., Mayer, M.P., Beisel, C., Klostermeier, D., Mogk, A., Reinstein, J. and Bukau, B. (1999) Proc. Natl. Acad. Sci. U.S.A. **96**, 5452–5457
22. Naylor, D.J., Hoogenraad, N.J., and Høj, P.B. (1996) FEBS Lett. **396**, 181–188
23. Naylor, D.J., Stines, A.P., Hoogenraad, N.J. and Høj, P.B. (1998) J. Biol. Chem. **273**, 21169–21177
24. Schlicher, T. and Soll, J. (1997) Plant Mol. Biol. **33**, 181–185
25. Mayer, M.P. and Bukau, B. (1998) Biol. Chem. **379**, 261–268
26. Zhu, X., Zhao, X., Burkholder, W.F., Gragerov, A., Ogata, C.M., Gottesman, M.E. and Hendrickson, W.A. (1996) Science **272**, 1606–1614
27. Flynn, G.C., Pohl, J., Flocco, M.T. and Rothman, J.E. (1991) Nature (London) **353**, 726–730
28. Landry, S.J., Jordan, R., McMacken, R. and Gierasch, L.M. (1992) Nature (London) **355**, 455–457
29. Blond-Elguindi, S., Cwirla, S.E., Dower, W.J., Lipshutz, R.J., Sprang, S.R., Sambrook, J.F. and Gething, M.-J.H. (1993) Cell **75**, 717–728
30. Fourie, A.M., Sambrook, J.F. and Gething, M.-J.H. (1994) J. Biol. Chem. **269**, 30470–30478
31. Gragerov, A., Zeng, L., Zhao, X., Burkholder, W. and Gottesman, M.E. (1994) J. Mol. Biol. **235**, 848–854
32. Rüdiger, S., Germeroth, L., Schneider-Mergener, J. and Bukau, B. (1997) EMBO J. **16**, 1501–1507
33. Rüdiger, S., Buchberger, A. and Bukau, B. (1997) Nat. Struct. Biol. **4**, 342–349
34. Fourie, A.M., Hupp, T.R., Lane, D.P., Sang, B.-C., Barbosa, M.S., Sambrook, J.F. and Gething, M.-J.H. (1997) J. Biol. Chem. **272**, 19471–19479
35. Liberek, K., Wall, D. and Georgopoulos, C. (1995) Proc. Natl. Acad. Sci. U.S.A. **92**, 6224–6228
36. Diamant, S. and Goloubinoff, P. (1998) Biochemistry **37**, 9688–9694
37. Pierpaoli, E.V., Sandmeier, E, Schönfeld, H.-J. and Christen, P. (1998) J. Biol. Chem. **273**, 6643–6649
38. Liberek, K., Marszalek, J., Ang, D., Georgopoulos, C. and Zylicz, M. (1991) Proc. Natl. Acad. Sci. U.S.A. **88**, 2874–2878
39. McCarty, J.S., Buchberger, A., Reinstein, J. and Bukau, B. (1995) J. Mol. Biol. **249**, 126–137
40. Gässler, C.S., Buchberger, A., Laufen, T., Mayer, M.P., Schröder, H., Valencia, A. and Bukau, B. (1998) Proc. Natl. Acad. Sci. U.S.A. **95**, 15229–15234
41. Liberek, K., Skowyra, D., Zylicz, M., Johnson, C. and Georgopoulos, C. (1991) J. Biol. Chem. **266**, 14491–14496
42. Schmid, D., Baici, A., Gehring, H. and Christen, P. (1994) Science **263**, 971–973
43. Palleros, D.R., Reid, K.L., Shi, L., Welch, W.J. and Fink, A.L (1993) Nature (London) **365**, 664–666

44. Palleros, D.R., Shi, L., Reid, K.L. and Fink, A.L (1994) J. Biol. Chem. **269**, 13107–13114
45. Greene, L., Zinner, R., Naficy, S. and Eisenberg, E. (1995) J. Biol. Chem. **270**, 2967–2973
46. Buchberger, A., Theyssen, H., Schröder, H., McCarty, J.S., Virgallita, G., Milkereit, P., Reinstein, J. and Bukau, B. (1995) J. Biol. Chem. **270**, 16903–16910
47. Szabo, A., Langer, T., Schröder, H., Flanagan, J., Bukau, B. and Hartl, F.-U. (1994) Proc. Natl. Acad. Sci. U.S.A. **91**, 10345–10349
48. Theyssen, H., Schuster, H.-P., Packschies, L., Bukau, B. and Reinstein, J. (1996) J. Mol. Biol. **263**, 657–670
49. Russell, R., Jordan, R. and McMacken, R. (1998) Biochemistry **37**, 596–607
50. Packschies, L., Theyssen, H., Buchberger, A., Bukau, B., Goody, R.S. and Reinstein, J. (1997) Biochemistry **36**, 3417–3422
51. Banecki, B and Zylicz, M. (1996) J. Biol. Chem. **271**, 6137–6143
52. Wawrzynów, A. and Zylicz, M. (1995) J. Biol. Chem. **270**, 19300–19306
53. Wawrzynów, A., Wojtkowiak, D., Marszalek, J., Banecki, B., Jonsen, M., Graves, B., Georgopoulos, C. and Zylicz, M. (1995) EMBO J. **14**, 1867–1877
54. Wickner, S., Hoskins, J. and McKenney, K. (1991) Nature (London) **350**, 165–167
55. Hoffmann, H.J., Lyman, S.K., Lu, C., Petit, M.-A. and Echols, H. (1992) Proc. Natl. Acad. Sci. U.S.A. **89**, 12108–12111
56. Gamer, J., Multhaup, G., Tomoyasu, T., McCarty, J.S., Rüdiger, S., Schönfeld, H.-J., Schirra, C., Bujard, H. and Bukau, B. (1996) EMBO J. **15**, 607–617
57. Liberek, K. and Georgopoulos, C. (1993) Proc. Natl. Acad. Sci. U.S.A. **90**, 11019–11023
58. Zylicz, M., Ang, D. and Georgopoulos, C. (1987) J. Biol. Chem. **262**, 17437–17442
59. Naylor, D.J., Ryan, M.T., Condron, R., Hoogenraad, N.J., and Høj, P.B. (1995) Biochim. Biophys. Acta **1248**, 75–79
60. Naylor, D.J., Stines, A.P., Hoogenraad, N.J. and Høj, P.B. (1998) J. Biol. Chem. **273**, 21169–21177
61. Bolliger, L., Deloche, O., Glick, B.S., Georgopoulos, C., Jenö, P., Kronidou, N., Horst, M., Morishima, N. and Schatz, G. (1994) EMBO J. **13**, 1998–2006
62. Mehta, A.D., Lee, J.Y. and Palter, K.B. (1997) in Guidebook to Molecular Chaperones and Protein-Folding Catalysts (Gething, M.-J., ed.), pp. 141–142, Oxford University Press, New York
63. Harrison, C.J., Hayer-Hartl, M., Di Liberto, M., Hartl, F.-U. and Kuriyan, J. (1997) Science **276**, 431–435
64. Osipiuk, J., Georgopoulos, C. and Zylicz, M. (1993) J. Biol. Chem. **268**, 4821–4827
65. Schönfeld, H.-J., Schmidt, D., Schröder, H. and Bukau, B. (1995) J. Biol. Chem. **270**, 2183–2189
66. Wu, B., Wawrzynow, A., Zylicz, M. and Georgopoulos, C. (1996) EMBO J. **15**, 4806–4816
67. Ziegelhoffer, T., Lopez-Buesa, P. and Craig, E. (1995) J. Biol. Chem. **270**, 10412–10419
68. Höhfeld, J., Minami, Y., and Hartl, F.-U. (1995) Cell **83**, 589–598
69. Minami, Y., Höhfeld, J., Ohtsuka, K. and Hartl, F.-U. (1996) J. Biol. Chem. **271**, 19617–19624
70. Höhfeld, J. and Jentsch, S. (1997) EMBO J. **16**, 6209–6216
71. Bimston, D., Song, J., Winchester, D., Takayama, S., Reed, J.C. and Morimoto, R.I. (1998) EMBO J. **17**, 6871–6878
72. Lüders, J., Demand, J., Schönfelder, S., Frien, M., Zimmermann, R. and Höhfeld, J. (1999) Biol. Chem. **379**, 1217–1226
73. Froesch, B.A., Takayama, S. and Reed, J.C. (1998) J. Biol. Chem. **273**, 11660–11666
74. Takayama, S., Xie, Z. and Reed, J.C. (1999) J. Biol. Chem. **274**, 781–786
75. Frydman, J. and Höhfeld, J. (1997) Trends Biochem. Sci. **22**, 87–92
76. Buchner, J. (1999) Trends Biochem. Sci. **24**, 136–141
77. Mayer, M.P. and Bukau, B. (1999) Curr. Biol. **9**, 322–325
78. Ellis, R.J. (1992) Nature (London) **358**, 191

79. Ryan, M.T., Naylor, D.J., Hoogenraad, N.J. and Høj, P.B. (1995) J. Biol. Chem. **270**, 22037–22043
80. Hunt, J.F., Weaver, A.J., Landry, S.J., Gierasch, L. and Deisenhofer, J. (1996) Nature (London) **379**, 37–45
81. Klumpp, M., Baumeister, W. and Essen, L.O. (1997) Cell **91**, 263–270
82. Ditzel, L., Löwe, J., Stock, D., Stetter, K.-O., Huber, H., Huber, R. and Steinbacher, S. (1998) Cell **93**, 125–138
83. Geissler, S., Siegers, K. and Schiebel, E. (1998) EMBO J. **17**, 952–966
84. Vainberg, I.E., Lewis, S.A., Rommelaere, H., Ampe, C., Vandekerckhove, J., Klein, H. and Cowan, N.J. (1998) Cell **93**, 863–873
85. Siegers, K., Waldmann, T., Leroux, M.R., Grein, K., Shevchenko, A., Schiebel, E. and Hartl, F.-U. (1998) EMBO J. **18**, 75–84
86. Hansen, W.J., Cowan, N.J. and Welch, W.J. (1999) J. Cell Biol. **145**, 265–277
87. Leroux, M.R., Fändrich, M., Klunker, D., Siegers, K., Lupas, A.N., Brown, J.R., Schiebel, E., Dobson, C. and Hartl, F.U. (1999) EMBO J. **18**, 6730–6743
88. Braig, K., Otwinowski, Z., Hedge, R., Boisvert, D.C., Joachimiak, A., Horwich, A.L. and Sigler, P.B. (1994) Nature (London) **371**, 578–586
89. Braig K., Adams, P.D. and Brünger, A.T. (1995) Nat. Struct. Biol. **2**, 1083–1094
90. Hunt, J.F., Weaver, A.J., Landry, S.J., Gierasch, L. and Deisenhofer, J. (1996) Nature (London) **379**, 37–45
91. Mande, S.C., Mehra, V., Bloom, B. and Hol, W.G. (1996) Science **271**, 203–207
92. Boisvert, D.C., Wang, J., Horwich, A.L. and Sigler, P.B. (1996) Nat. Struct. Biol. **3**, 170–177
93. Xu, Z., Horwich, A.L. and Sigler, P.B. (1997) Nature (London) **388**, 741–750
94. Landry, S.J., Zeilstra, R.J., Fayet, O., Georgopoulos, C. and Gierasch, L.M. (1993) Nature (London) **364**, 255–258
95. Fenton, W.A., Kashi, Y., Furtak, K. and Horwich, A.L. (1994) Nature (London) **371**, 614–619
96. Chen, S., Roseman, A.M., Hunter, A.S., Wood, S.P., Burston, S.G., Ranson, N.A., Clarke, A.R. and Saibil, H.R. (1994) Nature (London) **371**, 261–264
97. Thiyagarajan, P., Henderson, S.J. and Joachimiak, A. (1996) Structure **4**, 79–88
98. Chen, L. and Sigler, P.B. (1999) Cell **99**, 757–768
99. Langer, T., Pfeifer, G., Martin, J., Baumeister, W. and Hartl, F.U. (1992) EMBO J. **11**, 4757–4765
100. Martin, J., Mayhew, M., Langer, T. and Hartl, F.U. (1993) Nature (London) **366**, 228–233
101. Weissman J.S., Hohl, C.M., Kovalenko, O., Kashi, Y., Chen, S., Braig, K., Saibil, H.R., Fenton, W.A. and Horwich, A.L (1995) Cell **83**, 577–587
102. Weissman, J.S., Rye, H.S., Fenton, W.A., Beechem, J.M. and Horwich, A.L (1996) Cell **84**, 481–490
103. Hayer-Hartl, M.K., Weber, F. and Hartl, F.U. (1996) EMBO J. **15**, 6111–6121
104. Mayhew, M., da Silva, A.C.R., Martin, J., Erdjument-Bromage, H., Tempst, P. and Hartl, F.U. (1996) Nature (London) **379**, 420–426
105. Roseman, A., Chen, S., White, H., Braig, K. and Saibil, H. (1996) Cell **87**, 241–251
106. Ellis, R.J. (1994) Curr. Biol. **4**, 633–635
107. Ellis, R.J. (1996) Folding Des. **1**, R9–R15
108. Shtilerman, M., Lorimer, G.H. and Englander, W. (1999) Science **284**, 822–825
109. Saibil, H.R. (1994) Nat. Struct. Biol. **1**, 838–842
110. Todd, M.J., Viitanen, P.V. and Lorimer, G.H. (1994) Science **265**, 659–666
111. Rye, H.S., Burston, S.G., Fenton, W.A., Beechem, J.M., Xu, Z., Sigler, P.B. and Horwich, A.L. (1997) Nature (London) **388**, 792–798
112. Rye, H.S., Roseman, A.M., Chen, S., Furtak, K., Fenton, W.A., Saibil, H.R. and Horwich, A.L. (1999) Cell **97**, 325–338
113. Kad, N.M., Ranson, N.A., Cliff, M.J. and Clarke, A.R. (1998) J. Mol. Biol. **278**, 267–278

114. Lorimer, G. (1997) Nature (London) **388**, 720–723
115. Hayer-Hartl, M.K., Martin, J. and Hartl, F.U. (1995) Science **269**, 836–841
116. Burston, S.G., Ranson, N.A. and Clarke, A.R. (1995) J. Mol. Biol. **249**, 138–152
117. Todd, M.J., Lorimer, G.H. and Thirumalai, D. (1996) Proc. Natl. Acad. Sci. U.S.A. **93**, 4030–4035
118. Weissman, J.S., Kashi, Y., Fenton, W.A. and Horwich, A.L (1994) Cell **78**, 693–702
119. Smith, K.E. and Fisher, M.T. (1995) J. Biol. Chem. **270**, 21517–21523
120. Taguchi, H. and Yoshida, M. (1995) FEBS Lett. **359**, 195–198
121. Burston, S.G., Weissman, J.S., Farr, G.W., Fenton, W.A. and Horwich A.L. (1996) Nature (London) **383**, 96–99
122. Ranson, N.A., Burston, S.G. and Clarke, A.R. (1997) J. Mol. Biol. **266**, 656–664
123. Martin, J. and Hartl, F.U. (1997) Proc. Natl. Acad. Sci. U.S.A. **94**, 1107–1112
124. Viitaen, P.V., Gatenby, A.A. and Lorimer, G.H. (1992) Protein Sci. **1**, 363–369
125. Horwich, A.L., Low, K.B., Fenton, W.A., Hirschfield, I.N. and Furak, K. (1993) Cell **74**, 909–917
126. Ewalt, K.L., Hendrick, J.P., Houry, W.A. and Hartl, F.-U. (1997) Cell **90**, 491–500
127. Houry, W.A., Frishman, D., Eckerskorn, C., Lottspeich, F. and Hartl, F.-U. (1999) Nature (London) **402**, 147–154
128. Fayet, O., Ziegelhoffer, T. and Georgopoulos, C. (1989) J. Bacteriol. **171**, 1379–1385
129. Buckle, A.M., Zahn, R. and Fersht, A.R. (1997) Proc. Natl. Acad. Sci. U.S.A. **94**, 3571–3575
130. Llorca, O., McCormack, E.A., Hynes, G., Grantham, J., Cordell, J., Carrascosa, J.L., Willison, K.R., Fernandez, J.J. and Valpuesta, J.M. (1999) Nature (London) **402**, 693–696
131. Thulasiraman, V., Yang, C.-F. and Frydman, J. (1999) EMBO J. **18**, 85–95
132. Melki, R. and Cowan, N.J. (1994) Mol. Cell. Biol. **14**, 2895–2904
133. Johnson, J.L. and Craig, E.A. (1997) Cell **90**, 201–204
134. Nathan, D.F., Vos, M.H. and Lindquist, S. (1997) Proc. Natl. Acad. Sci. U.S.A. **94**, 12949–12956
135. Das, B., Chattopadhyay, S., Bera, A.K. and Dasgupta, C. (1996) Eur. J. Biochem. **235**, 613–621
136. Kudlicki, W., Coffman, A., Kramer, G. and Hardesty, B. (1997) Folding Des. **2**, 101–108
137. Ellis, R.J. (1997) Biochem. Biophys. Res. Commun. **238**, 687–692
138. Caldas, T.D., Yaagoubi, A.E. and Richarme, G. (1998) J. Biol. Chem. **273**, 11478–11482
139. Teter, S.A., Houry, W.A., Ang, D., Tradler, T., Rockabrand, D., Fischer, G., Blum, P., Georgopoulos, C. and Hartl, F.U. (1999) Cell **97**, 755–765
140. Deuerling, E., Schulze-Specking, A., Tomoyasu, T., Mogk, A. and Bukau, B. (1999) Nature (London) **400**, 693–696
141. Manning-Krieg, U.C., Scherer, P.E. and Schatz, G. (1991) EMBO J. **10**, 3273–3280
142. Mizzen, L.A., Kabiling, A.N. and Welch, W.J. (1991) Cell Regul. **2**, 165–179
143. Gaitanaris, G.A., Vysokanov, A., Hung, S.-C., Gottesman, M.E. and Gragerov, A. (1994) Mol. Microbiol. **14**, 861–869
144. Saijo, T., Welch, W.J. and Tanaka, K. (1994) J. Biol. Chem. **269**, 4401–4408
145. Rospert, S., Looser R., Dubaquie, Y., Matouschek, A., Glick, B.S. and Schatz, G. (1996) EMBO J. **15**, 764–774
146. Frydman, J., Nimmesgern, E., Ohtsuka, K. and Hartl, F.-U. (1994) Nature (London) **370**, 111–117
147. Frydman, J. and Hartl, F.-U. (1996) Science **272**, 1497–1502
148. Heyrovshá, N., Frydman, J., Höhfeld, J. and Hartl, F.-U. (1998) Biol. Chem. **379**, 301–309
149. Beckmann, R.P., Mizzen, L.A. and Welch, W.J. (1990) Science **248**, 850–854
150. Nelson, R.J., Ziegelhoffer, T., Nicolet, C., Werner-Washburne, M. and Craig, E.A. (1992) Cell **71**, 97–105
151. Hansen, W.J., Lingappa, V.R. and Welch, W.J. (1994) J. Biol. Chem. **269**, 26610–26613
152. Eggers, D.K., Welch, W.J. and Hanson, W.J. (1997) Mol. Biol. Cell **8**, 1559–1573

153. Pfund, C., Lopez-Hoyo, N., Ziegelhoffer, T., Schilke, B.A., Lopez-Buesa, P., Walter, W.A., Weidmann, M. and Craig, E.A. (1998) EMBO J. **17**, 3981–3989
154. Langer, T., Lu, C., Echols, H., Flanagen, J., Hayer, M.K. and Hartl, F.U. (1992) Nature (London) **356**, 683–689
155. Csermely, P., Schnaider, T., Sóti, C., Prohászka, Z. and Nardai, G. (1998) Pharmacol. Ther. **79**, 129–168
156. Buchberger, A., Schröder, H., Hesterkamp, T., Schönfeld, H.-J. and Bukau, B. (1996) J. Mol. Biol. **261**, 328–333
157. Bukau, B., Hesterkamp, T. and Luirink, J. (1996) Trends Cell Biol. **6**, 480–486
158. Ellis, R.J. (1999) Curr. Biol. **9**, 137–139
159. Glover, J.R. and Lindquist, S. (1998) Cell **94**, 73–82

Biochem. Soc. Symp. **68**, 69–82
(Printed in Great Britain)

5

Defining the structure of the substrate-free state of the DnaK molecular chaperone

Joanna F. Swain*, Renuka Sivendran‡ and Lila M. Gierasch*†[1]

*Department of Biochemistry and Molecular Biology, University of Massachusetts, Amherst, MA 01003, U.S.A., †Department of Chemistry, University of Massachusetts, Amherst, MA 01003, U.S.A., and ‡Molecular and Cellular Biology Graduate Program, University of Massachusetts, Amherst, MA 01003, U.S.A.

Abstract

Members of the Hsp70 (heat-shock protein of 70 kDa) family of molecular chaperones bind to exposed hydrophobic stretches on substrate proteins in order to dissociate molecular complexes and prevent aggregation in the cell. Substrate affinity for the C-terminal domain of the Hsp70 is regulated by ATP binding to the N-terminal domain utilizing an allosteric mechanism. Our multi-dimensional NMR studies of a substrate-binding domain fragment (amino acids 387–552) from an *Escherichia coli* Hsp70, DnaK(387–552), have uncovered a pH-dependent conformational change, which we propose to be relevant for the full-length protein also. At pH 7, the C-terminus of DnaK(387–552) mimics substrate by binding to its own substrate-binding site, as has been observed previously for truncated Hsp70 constructs. At pH 5, the C-terminus is released from the binding site, such that DnaK is in the substrate-free state 10–20% of the time. We propose that the mechanism for the release of the tail is a loss of affinity for substrate at low pH. The pH-dependent fluorescence changes at a tryptophan residue near the substrate-binding pocket in full-length DnaK lead us to extend these conclusions to the full-length DnaK as well. In the context of the DnaK substrate-binding domain fragment, the release of the C-terminus from the substrate-binding site provides our first glimpse of the empty conformation of an Hsp70 substrate-binding domain containing a portion of the helical subdomain.

[1]To whom correspondence should be addressed at the Department of Biochemistry and Molecular Biology.

Introduction

Much effort in recent years has been focused on understanding the mechanism of allostery in the Hsp70 (heat-shock protein of 70 kDa) family of molecular chaperones. Members of this family have a variety of functions, which include preventing protein aggregation during periods of cellular stress, assisting nascent protein chains both to translocate into organelles and to fold, preparing proteins for proteolytic digestion, and dissociating molecular complexes, such as the clathrin coat [1,2]. In order to carry out these functions, Hsp70 molecules bind to exposed hydrophobic sequences on substrate proteins, utilizing their C-terminal domains. The affinity of this interaction is regulated by nucleotide binding to the N-terminal domain of Hsp70; in the ADP-bound state, substrates bind with high affinity and slow on/off kinetics, but when ADP is exchanged for ATP, substrate affinity drops and the on/off rates for the interaction increase. How information regarding binding site occupancy is transmitted from the nucleotide-binding domain to the substrate-binding domain, and vice versa, is as yet unknown, although recent mutational and structural analyses have allowed several models to be formulated [2–6].

The lack of a high-resolution structure of an intact Hsp70 protein has made it difficult to characterize the allosteric mechanism in structural terms. However, several crystal and solution structures for isolated domains have been determined. Crystal structures of the ATPase domain from the bovine cytosolic chaperone, Hsc70, as well as that from the *Escherichia coli* Hsp70, DnaK, in complex with its exchange factor, GrpE, portray a globular mixed α/β-domain with a deep cleft containing the ATP-binding site [7,8]. Based on structural similarity with hexokinase and actin, the two lobes of the nucleotide-binding domain have been proposed to hinge inward and outward during the cycle of ATP binding and hydrolysis, a motion that might propagate signals to the substrate-binding domain. In the intact protein, the ATPase domain is connected via a short proteolytically sensitive linker to the substrate-binding domain. Small angle X-ray scattering studies [9,10], as well as the demonstration that an ATP-induced change in the fluorescence of an N-terminal tryptophan residue requires the presence of the substrate-binding domain [11], argue that the two domains dock in a stable manner when ATP binds.

The substrate-binding domain, on the other hand, consists of a β-subdomain containing two four-stranded anti-parallel β-sheets followed by an extended helix that caps the site of substrate binding and terminates in a helical bundle [3]. A hydrophobic heptapeptide substrate (NR; amino acid sequence Asn-Arg-Leu-Leu-Leu-Thr-Gly), co-crystallized in this DnaK substrate-binding domain (residues 389–607), binds in an extended conformation and is contacted primarily by residues in loops of the β-subdomain. The authors of this study proposed that the helical region is intrinsic to the mechanism of substrate release, based on a slight divergence in the orientation of the C-terminal end of helix αB in two different crystalline forms. In their model, they propose that ATP binding to the N-terminal domain favours formation of a hinge at residues 536–538, exposing the substrate-binding site in order to allow egress of substrate. However, we have shown recently that a severely truncated form of DnaK, lacking the entire helical subdomain, termed DnaK(1–507) retains the ability to support bacteriophage-λ replication *in vivo* when expressed in a DnaK-deficient strain [6]. Furthermore, DnaK(1–507) exhibits ATP-induced substrate release and peptide-induced ATPase activation, indicating that the helical subdomain is not crucial for interdomain communication in DnaK.

Attempts to study the substrate-binding domain in the empty conformation by NMR or crystallography have been stymied for several years by oligomerization and self-binding behaviour of these domains in the absence of added substrates. NMR solution structures of substrate-binding domains of DnaK and Hsc70 that were truncated in the middle of the helical subdomain (corresponding to residues 386–561 and 383–540 respectively) revealed a surprising intra-molecular interaction in which C-terminal tail residues were bound tightly in the substrate-binding site, effectively mimicking an unfolded substrate [4,5]. In light of evidence that the helical region is not required for allostery in DnaK, the solution structure of an empty DnaK substrate-binding domain lacking all helical residues (corresponding to residues 393–507) has recently been solved [6]. In the absence of added substrate peptide, DnaK-(393–507) exhibits substantial conformational flexibility, especially in the C-terminal end of strand β3 and loops $L_{3,4}$ and $L_{5,6}$. Since addition of substrate peptide stabilized the structure of the β-domain, the structural differences observed were attributed to absence of bound substrate, rather than absence of the helical subdomain. Residues from loops $L_{1,2}$ and $L_{3,4}$ partially occlude the site occupied by substrate in previous structures, and the C-terminal half of strand β3 does not participate in a stable β-sheet interaction, as it does in the substrate-bound structures. The conformational differences in the absence of substrate led to a model for allosteric communication in which information about substrate-binding-site occupancy is transmitted via conformational changes in strand β3 to the region of the substrate-binding domain containing the N-terminal linker from the ATPase domain.

This model is consistent with previous biochemical studies relating the region near the N-terminus of strand β3 in the substrate-binding domain with allosteric function. Our random mutagenesis of the substrate-binding domain of DnaK identified a single point mutation, Lys-414→Ile, which abolishes function *in vivo* and demonstrates a complete lack of interdomain communication when assayed *in vitro* [12]. This site is of special interest because the proteolytic sensitivity of the Lys-414–Asn-415 peptide bond is increased substantially in the presence of ATP [11]. Furthermore, mutation of a conserved proline (Pro-419), in both DnaK and the mitochondrial Hsp70, leads to defects *in vivo* despite adequate substrate-binding capability [13,14]. These studies implicate loop $L_{2,3}$ in transduction of allosteric communication between the substrate-binding and ATPase domains of the Hsp70 proteins.

The structural data from studies of the isolated substrate-binding domain suggest that it is only marginally stable in the absence of substrate. Furthermore, the empty species will bind to any unfolded hydrophobic sequences available, resulting in oligomerization, or the intra-molecular binding observed in several NMR structures. However, we have discovered recently that a DnaK substrate-binding domain construct containing a large portion of the helical subdomain (corresponding to residues 387–552), undergoes a pH-dependent conformational change that results in release of the C-terminal tail residues from the binding pocket, allowing us to characterize the contribution of the helix to the stability and structure of the empty-site form.

Materials and methods

Preparation of DnaK(387–552)

DnaK(387–552) with an N-terminal histidine tag was expressed from plasmid pRLM212, provided by Roger McMacken (Johns Hopkins University, Baltimore, MD, U.S.A.), in *E. coli* BL21(λDE3). Cells were grown in ^{15}N-supplemented M9 minimal media to mid-exponential phase in shaker culture at 30°C, followed by a rapid shift to 42°C and an additional 5 h of shaking to induce protein expression. Cells were harvested by centrifugation at 1500 *g* for 30 min, and snap-frozen in liquid nitrogen. Thawed cells were resuspended in Buffer N [50 mM Na_2HPO_4 (pH 8.0), 0.5 M NaCl] and treated with lysozyme (1 mg/ml). After sonication, the soluble fraction was isolated by centrifugation at 27000 *g* for 30 min. The soluble extract was loaded on a Ni^{2+}-nitrilotriacetate agarose affinity column (Qiagen, Valencia, CA) equilibrated in Buffer N, and eluted with a gradient of imidazole up to a final concentration of 200 mM. Pooled fractions were concentrated in an ultrafiltration apparatus (Millipore Corp, Bedford, MA) and dialysed extensively against 20 mM sodium acetate (pH 5.0).

Preparation of NR peptide

NR peptide was synthesized in-house on an automated synthesizer using Fmoc (fluoren-9-ylmethoxy-carbonyl) chemistry (PE Biosystems, Foster City, CA). After purification by reverse-phase HPLC, the purity, concentration and molecular mass of the peptide were verified by analytical HPLC, quantitative amino acid analysis and mass spectrometry, respectively.

NMR sample preparation

[^{15}N]DnaK(387–552) was extensively buffer-exchanged to 20 mM sodium acetate-*d3* (pH 5.0) or 20 mM Na_2HPO_4 (pH 7.0) using a microconcentration device (Millipore). Protein samples at 0.4–0.8 mM were supplemented with 10% ^2H_2O, 0.02% NaN_3, and 0.1 mM 3-(trimethylsilyl) propane sulphonic acid. For peptide titration, freeze-dried NR peptide was added to 0.4 mM [^{15}N]DnaK(387–552) to a final concentration of 5.8 mM.

NMR experiments

Standard gradient $^1H,^{15}N$-HSQC (heteronuclear single-quantum coherence spectroscopy), $^1H,^{15}N$-TOCSY-HSQC and $^1H,^{15}N$-NOESY-HSQC experiments [15] were collected on a 500 MHz Bruker AMX spectrometer. Data were processed using Felix97, and spectra were analysed using the program XEASY [16]. To determine the fraction of DnaK(387–552) in the empty conformation at pH 5.0, the integrals of sequential NOESY crosspeaks between the amide protons of Asn-537 and Gln-538, and between those of

Asp-540 and His-541 were compared with the integrals expected if the C-terminal sequence were bound in the binding site or, alternatively, if the C-terminus formed a helical structure. Thus, the mean distance between each pair of amide protons in the family of 25 NMR structures of DnaK(386–561) (3.82 Å for 537–538, and 4.33 Å for 540–541 [4]) was used to calculate the integral expected with DnaK(387–552) in the fully tail-bound conformation. In order to accomplish this, the mean integral of the Hα(i) to HN(i+1) NOESY crosspeaks from well-dispersed sequential β-sheet residues was used to generate a conversion factor between the sixth power of the interproton radius and the integral magnitude for the ^{1}H,^{15}N-NOESY-HSQC spectrum. Similarly, the typical distance between sequential protons in α-helix (2.8 Å [17]) was used to calculate the expected integral magnitude if the C-terminal residues of DnaK(387–552) formed an α-helix. Finally, assuming that the protein only samples these two states, the fraction (x) in helical conformation was calculated according to $I_m = (1 - x)I_t + (x)I_h$, where I_m is the measured integral, I_t is the expected integral for the tail-bound state, and I_h is the expected integral for the helical state.

Fluorescence experiments

The mutant of DnaK used for fluorescence studies (Ala-429→Trp/Trp-102→Phe/Glu-551→Cys) contains a single tryptophan in the substrate-binding domain, and was created by site-directed mutagenesis with quick-change PCR. Protein was expressed and purified as described for wild-type DnaK [12]. The fluorescence emission spectra of 4 μM protein in either buffer consisting of 20 mM *N*-[2-hydroxyethyl]piperazine-*N*′-[2-ethansulphonic acid], 100 mM KCl and 5 mM $MgCl_2$ (pH 7.0 and 8.0), or buffer consisting of 20 mM potassium acetate, 80 mM KCl and 5 mM $MgCl_2$ (pH 5.0 and 6.0), were measured using a fluorimeter (Photon Technology International, Lawrenceville, NJ) with excitation at 295 nm.

Results

The DnaK substrate-binding domain construct used for structural studies extends from residue 387, located in the linker region between the N- and C-terminal domains, to residue 552, which occurs at the end of helix αB in the NR-bound crystal structure. The pattern of ^{1}H,^{15}N-HSQC peaks in an experiment collected at pH 7.0 and 35°C correlates strongly with the backbone amide assignments from a slightly larger substrate-binding domain fragment, DnaK(386–561), at 25°C (Figure 1a) [4]. The slight peak-shifts observed are the result of temperature effects, as a spectrum collected at 25°C overlays completely with the assigned peak positions. These data argue that DnaK(387–552) forms a structure at pH 7.0 in which residues near the C-terminus are bound in the substrate-binding site, like the structure that was determined for DnaK-(386–561) [4]. However, significant peak shifts are evident when the pH is lowered to 5.0 (Figure 1b), indicating a pH-dependent conformational change in DnaK(387–552). We propose that this conformational change is initiated by protonation of histidine residues, and results ultimately in the release of the sub-

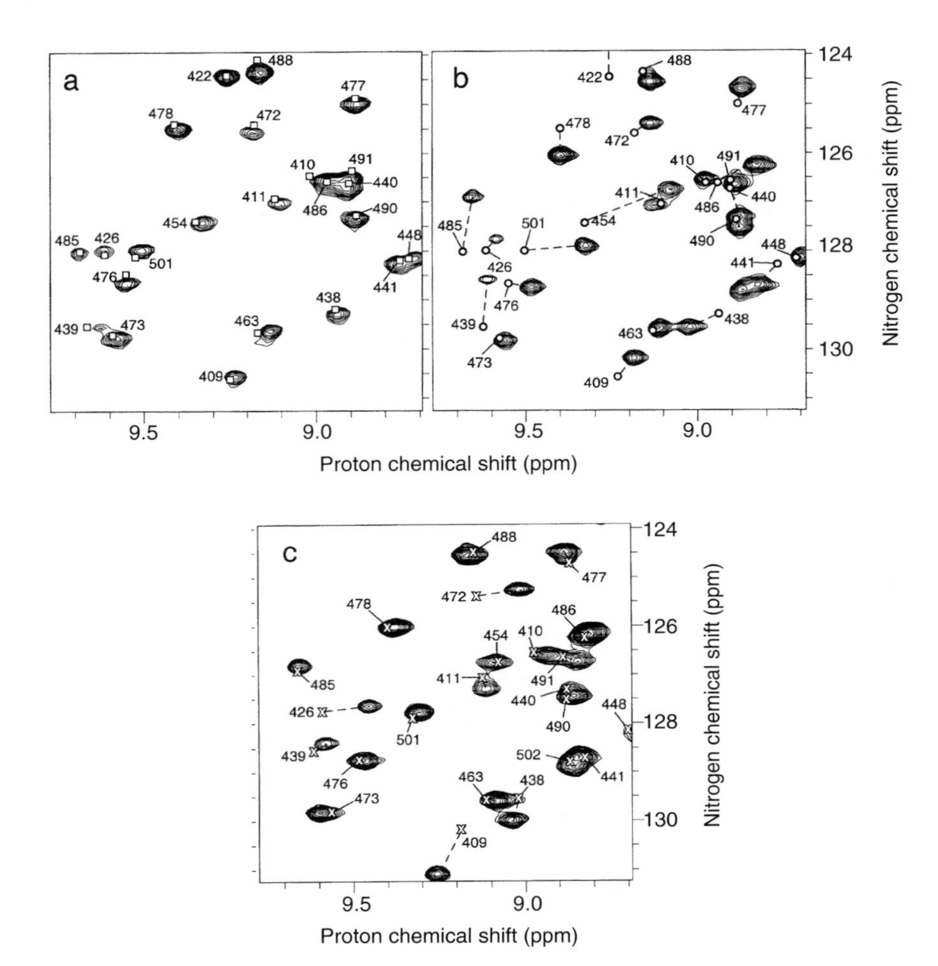

Figure 1 A pH-dependent conformational change in DnaK(387–552). $^{1}H,^{15}N$-HSQC spectra of 0.4 mM ^{15}N-labelled DnaK(387–552) collected at 35°C are shown. In (a), the HSQC at pH 7.0 is overlaid with squares indicating the peak positions for the DnaK(386–561) fragment at pH 7.0 and 25°C [4], labelled according to the residue number. In (b), the HSQC spectrum for DnaK (387–552) at pH 5.0 is overlaid with circles indicating the location of its peaks at pH 7.0. Dashed lines connect the pH 7.0 peak positions with the corresponding pH 5.0 peaks. In (c), the HSQC of DnaK(387–552) with 5.8 mM NR peptide at pH 5.0 is shown, with the peak positions at this pH in the absence of NR indicated by ×. Dashed lines correlate peak positions in the absence of NR with NR-bound peaks. Note that since 5.8 mM NR is not sufficient to fully saturate DnaK(387–552), small peaks representing the NR-free state are still observable in this spectrum.

strate-mimicking C-terminus from the substrate-binding site. Consistent with this hypothesis, DnaK(387–552) is capable of binding the NR peptide at pH 5.0

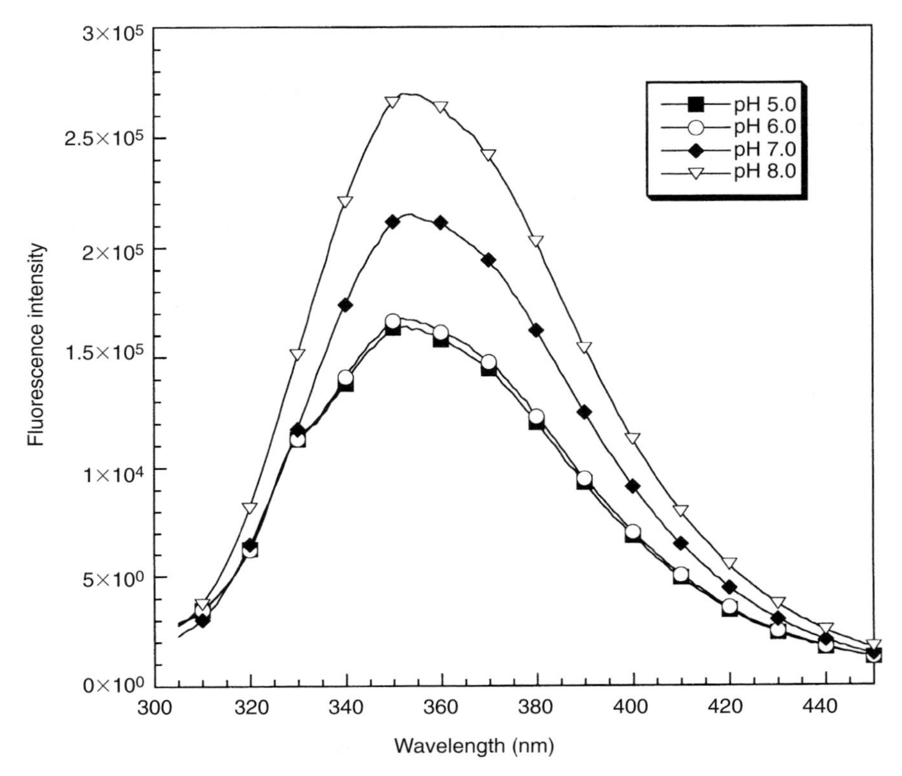

Figure 2 pH-dependence of tryptophan fluorescence in full-length DnaK. Fluorescence emission spectra of a mutant of DnaK (Ala-429→Trp/Trp-102→Phe/Glu-551→Cys) that has a single tryptophan in the peptide-binding domain; spectra at pH 5.0, 6.0, 7.0 and 8.0 are shown.

(Figure 1c). These data are in contrast to previous unsuccessful efforts to demonstrate peptide binding to the DnaK(386–561) fragment at pH 7.0 [4]. As seen in Figure 1(c), only a subset of resonances is affected by NR binding, indicating that NR is binding to a specific site on the protein. We conclude that the DnaK(387–552) substrate-binding domain construct experiences a pH-dependent conformational change; at pH 7.0, the binding site is filled by a sequence near the C-terminus, but, as the pH is reduced, the substrate-binding domain increasingly assumes a conformation in which the binding site is empty.

These data suggest that the affinity of the DnaK substrate-binding domain for the C-terminal substrate-mimicking sequence is dependent on the charge state of histidine residues. One possibility is that the two histidines (residues 541 and 544) within the sequence that binds intra-molecularly to the binding site become protonated between pH 7.0 and 5.0, resulting in a significant decrease in hydrophobic character and concomitant loss of binding affinity. However, we have evidence that a pH-dependent conformational change occurs also in full-length DnaK, and may provide additional impetus to the loss of affinity for the tail residues. Our reporter for this conformational change is the fluorescence of a tryptophan mutant in full-length DnaK. In other structures of DnaK, Ala-429 in loop $L_{3,4}$ forms an arch over the bound substrate with the aid of Met-404 on loop $L_{1,2}$ [3,4,6]. The fluorescence characteristics of a tryptophan residue engineered at position 429 report on environmental changes near the substrate-binding pocket. As seen in Figure 2, the quantum yield of Trp-429 fluorescence is exquisitely sensitive to pH, indicating a shift to a less hydrophobic environment below pH 6.0. Interestingly, a recent report describes a significant increase in the off-rate for a substrate peptide bound to full-length DnaK below pH 6.7 [18]. Thus, an alternative explanation for the release of the substrate-mimicking tail in the context of the DnaK(387–552) fragment is that the substrate-binding affinity of full-length DnaK is pH-dependent.

In order to study the empty conformation of DnaK(387–552) at pH 5.0 by NMR, we assigned the $^{1}H,^{15}N$-HSQC peaks to particular backbone amides in the protein. Fortunately, the conformational changes that occur between pH 7.0 and 5.0 are in the fast-exchange regime, meaning that as the conformational equilibrium is shifted, the peaks shift incrementally according to the weighted population difference. Thus we were able to transfer pH 7.0 assignments for the longer DnaK(386–561) fragment (kindly provided by Erik Zuiderweg, University of Michigan, Ann Arbor, MI [4]) to the peaks in the pH 5.0 $^{1}H,^{15}N$-HSQC by incrementally lowering the pH and following the peaks as they shifted (Figure 3). In cases of spectral overlap, the accuracy of final sequential assignments was evaluated with the aid of three-dimensional $^{1}H,^{15}N$-TOCSY-HSQC experiments to verify spin systems, and a $^{1}H,^{15}N$-NOESY-HSQC experiment to determine sequential NOE connectivities.

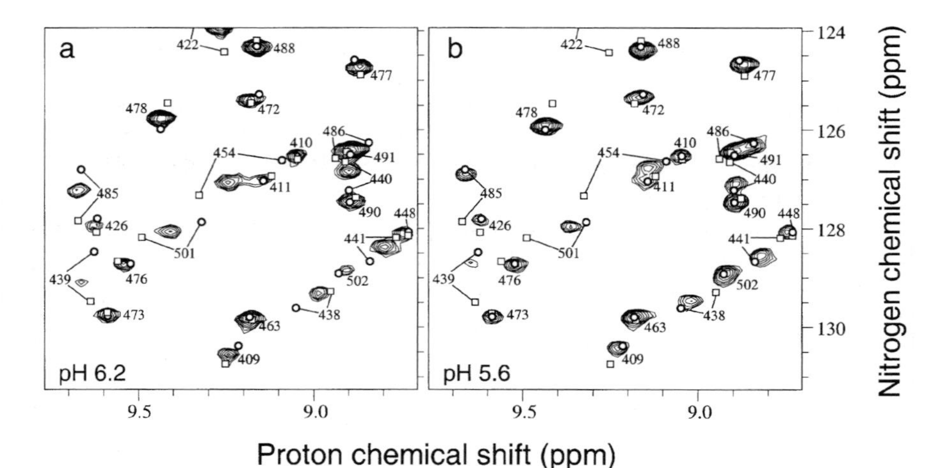

Figure 3 Fast-exchange kinetics of the pH-dependent conformational change. $^{1}H,^{15}N$-HSQC spectra of ^{15}N-labelled DnaK(387–552) collected at 25°C at pH 6.2 (a) and 5.6 (b), illustrating the incremental shift of amide peaks from pH 7.0 positions (indicated by squares) to pH 5.0 positions (circles) as the pH is reduced.

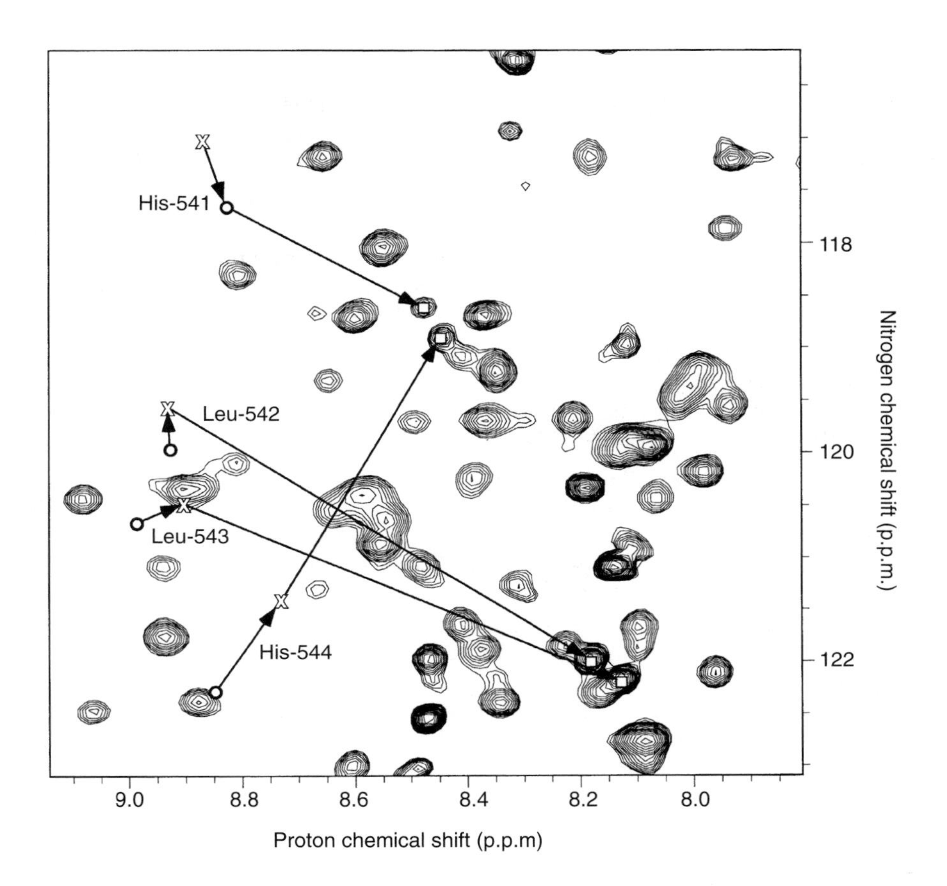

Figure 4 Chemical shift changes for the His-541–H-544 sequence. A $^{1}H,^{15}N$-HSQC spectrum of 0.4 mM ^{15}N-labelled DnaK(387–552) with 5.8 mM NR peptide, collected at 35°C and pH 5.0, is shown. Circles, × and squares indicate peak positions at pH 7.0, pH 5.0, and pH 5.0 in the NR-bound state respectively for C-terminal tail residues His-541, Leu-542, Leu-543 and His-544.

It is interesting to note that binding of the peptide NR to DnaK(387–552) at pH 5.0 occurs in the slow-exchange regime, wherein titration of NR causes intensity loss at certain amide peaks with a concomitant appearance of new peaks at a slightly offset position, similar to what was observed for NR binding to the β-domain alone [6]. This behaviour contrasts with the kinetics of binding of the C-terminal tail, which are, apparently, much faster. These observations provide insight into the entropic advantage of binding a tethered ligand compared with a free ligand [19]. Because the C-terminal sequence is covalently linked to the substrate-binding domain fragment, it has less rotational and translational entropy to lose upon binding than would a free peptide. The cost of restricting the conformational space of the tail is reduced compared with that of NR, ultimately conferring an entropic enhancement to its on-rate.

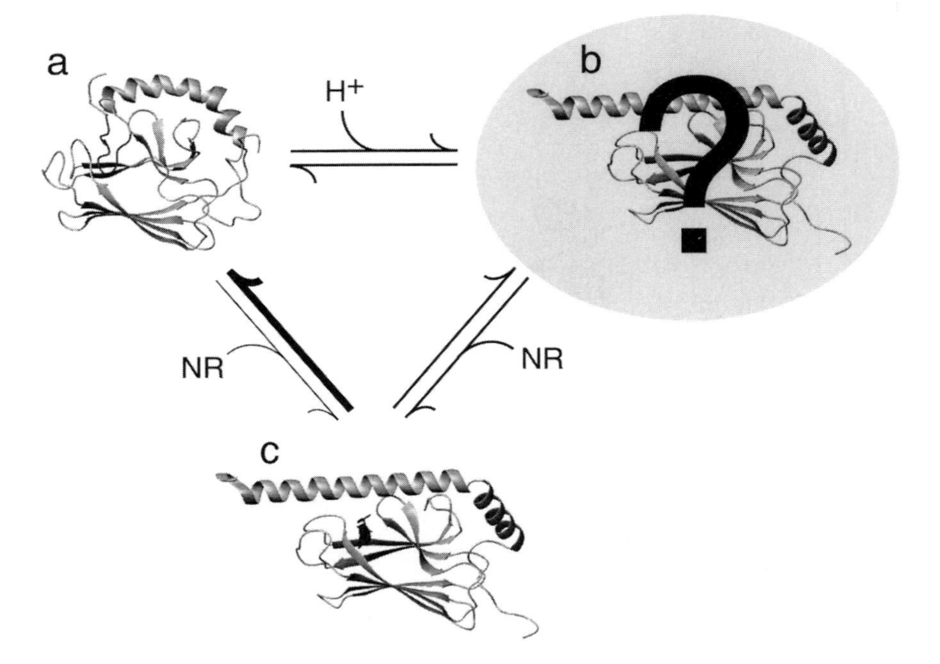

Figure 5 A model for the conformational equilibria observed in the DnaK(387–552) fragment. At pH 7.0, DnaK(387–552) forms a structure in which residues from the C-terminus bind in the substrate-binding site (a) similar to that of DnaK(386–561) {Protein Data Bank (PDB) accession code 2BPR [4]}. As the pH is reduced, an equilibrium is established between the tail-bound conformation and a, previously, uncharacterized state in which the substrate-binding site is empty. The latter form is represented schematically in (b). At pH 5.0, we estimate that 10–20% of DnaK(387–552) exists in an empty-site conformation. Addition of NR peptide shifts DnaK(387–552) to a third state, illustrated in (c), which has been characterized by X-ray crystallography (PDB accession code 1DKZ [3]). The figure was created with the aid of the molecular graphics program, MOLMOL [22].

The fact that DnaK(387–552) binds NR peptide at pH 5.0 suggests that the protein assumes an empty-site form in the absence of added substrate. Several lines of evidence, however, lead us to conclude that, at pH 5.0, we are in fact observing a equilibrium between the tail-bound conformation and an empty conformation. Foremost among these are significant spectral changes that occur in the C-terminal residues His541–His544 in the presence of NR peptide. When NR binds, the peaks for these residues shift further than for any other residue in the protein (Figure 4); these shifts are accompanied by substantial peak narrowing (from approx. 23 Hz in the absence of NR peptide to approx. 18 Hz in the NR-bound state). The spectral changes observed are surprising, since this sequence does not contact the peptide in the crystal structure of a substrate-binding domain complexed with NR [3]. However, in the NMR structure of DnaK(386–561), these are the residues that associate intra-molecu-

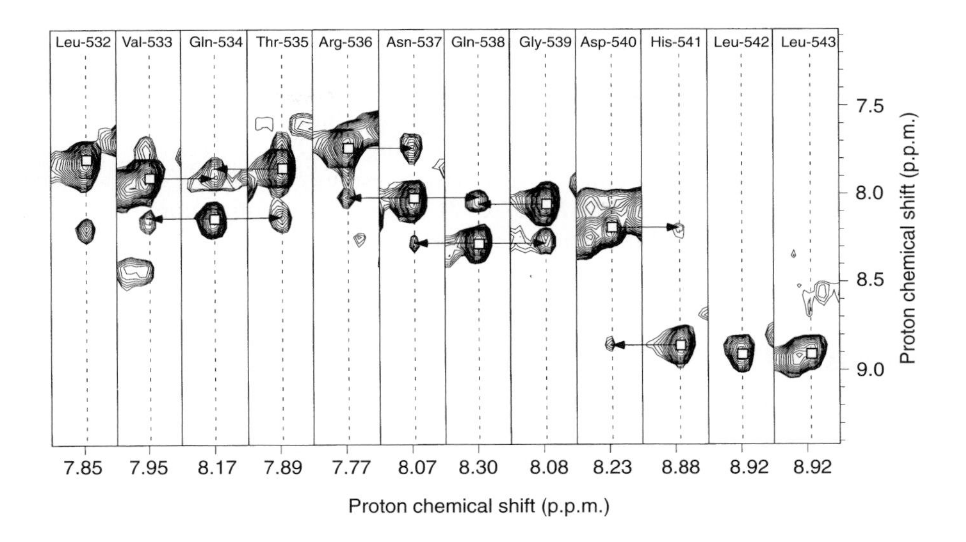

Figure 6 Estimation of population of an empty-state form. Slices shown are taken from a three-dimensional ^{1}H,^{15}N-NOESY-HSQC collected on 0.8 mM ^{15}N-labelled DnaK(387–552) at pH 5.0 and 35°C. Crosspeaks representing sequential amide-to-amide NOEs are indicated by arrows (squares indicate amide diagonal peaks). NOEs are typically only observed between protons that are less than 5 Å apart, so sequential amide NOEs are indicative of helical or turn sequences, and are not observed in extended chains. See the text for further details.

larly with the binding site [4]. Thus the most likely explanation for this behaviour is that, in the absence of NR, these residues switch between one state, in which they are bound in the substrate-binding site, and another, in which they are not; when NR binds, the equilibrium is shifted to favour the latter state (Figure 5).

Quantifying the fraction of the population that is in the empty conformation at pH 5.0 is a challenging prospect. The chemical shift changes observed between pH 7.0 and pH 5.0 result from a combination of the proposed conformational change, as well as changes in local electrostatics due to protonation of histidines and, possibly, some glutamate and aspartate residues. As such, the titration curves of most resonances do not saturate, and many have a complex shape (results not shown). Nevertheless, spectral evidence corroborates the proposal that the empty conformation is significantly populated. Sequential d_{NN} NOEs, the hallmark of helical structure, are observed up to, and including, residue His541 in a three-dimensional ^{1}H,^{15}N-NOESY-HSQC collected on DnaK(387–552) at pH 5.0 (Figure 6). Since the amide proton chemical shifts of residues 541–543 are degenerate, we cannot determine, on the basis of these data, whether helical character continues beyond residue 541. Nevertheless, the NOEs observed are not consistent with sequential amide-to-amide distances measured in the 25 NMR structures determined for DnaK-(386–561), in which C-terminal sequences were bound in the substrate-binding site [4]. In those structures, helix αB melts at Gln-534, after which the polypeptide chain loops around to allow residues Asp-539 to Arg-547 to bind in an extended conformation in the substrate-binding site. Assuming that we are observing a two-state equilibrium between a conformation, in which the C-terminal residues are bound in the substrate-binding site (Figure 5a) and an empty conformation in which the helix continues to residue 541 (Figure 5b), we can estimate that 10–20% of DnaK(387–552) exists in an empty conformation, based on the strength of the sequential amide to amide NOESY crosspeaks for residues 537–538 and 540–541 (see Materials and methods).

Discussion

This study of a substrate-binding domain fragment of the DnaK molecular chaperone has led to a more complete understanding of environmental influences on substrate binding to DnaK. In the context of the DnaK(387–552) fragment, we observe release of the substrate-mimicking C-terminus as the pH is lowered from 7.0 to 5.0. We propose that the conformational change that occurs in the fragment, i.e. release of the substrate-mimicking tail, is informing us about the pH-dependence of the substrate-binding affinity in full-length DnaK. A recent report that the off-rate of a substrate peptide bound to full-length DnaK is substantially increased below pH 6.7 corroborates the idea of a pH-dependent shift in substrate affinity [18]. Furthermore, our finding that a tryptophan residue engineered near the substrate-binding site in full-length DnaK shifts to a more hydrophilic environment between pH 8.0 and 5.0 suggests increased solvent exposure of the binding site at low pH. Several models describing the switch of the DnaK substrate-binding domain from a high-affinity state to a low-affinity state have invoked an increase in solvent accessibility of residues around the binding site [2,3].

These data raise the question of whether the pH sensitivity of the substrate-binding site has physiological relevance for the function of DnaK *in vivo*. The *E. coli* growth rate is related strongly to the pH of the growth medium: they grow optimally in the pH 6.0–7.0 range, but growth is strongly inhibited below pH 4.4 [20]. Acid shock in *E. coli* results in induction of DnaK protein expression, suggesting that DnaK plays a role in survival of acid shock similar to its role in heat-shock stress [21]. Thus pH regulation of DnaK substrate-binding affinity may indeed have important ramifications for maintenance of viability in acid-shock conditions.

Another interesting implication from our study of the conformational equilibria in the DnaK(387–552) fragment is that the DnaK substrate-binding domain is extremely difficult to observe in the empty conformation: attempts to crystallize the DnaK(389–607) fragment in the absence of substrate were frustrated by aggregation, DnaK(386–561) and Hsc70(385–543) fragments bind tightly to their own C-termini, and the helix-less DnaK(393–507) in the absence of substrate is characterized by loss of some structural integrity and regions of extreme mobility [3–6]. Perhaps these data are telling us something about the behaviour of Hsp70 substrate-binding domains *in vivo*. If these domains are indeed highly unstable in the empty conformation in the context of full-length DnaK, then we might predict that they do not spend much time in that state, or that interactions with other proteins or co-chaperones must occur for the empty state to be stable.

Finally, these studies have provided us with the exciting revelation that we can now study the empty conformation of a DnaK substrate-binding domain fragment containing a large portion of the helical region using NMR. Reducing the pH opens a window on the empty state of this construct which will allow us to characterize the contribution of the helix to the conformation and stability of the substrate-binding domain.

This work was supported by NIH grants R01 GM27616 to L.M.G. and F32 GM19573 to J.L.F. We gratefully acknowledge Roger McMacken (Johns Hopkins University) for the generous gift of the pRLM212 plasmid, and Erik Zuiderweg (University of Michigan) for providing us with the backbone chemical shifts for DnaK(386–561).

References

1. Hendrick, J.P. and Hartl, F.U. (1993) Annu. Rev. Biochem. **62**, 349–384
2. Rüdiger, S., Buchberger, A. and Bukau, B. (1997) Nat. Struct. Biol. **4**, 342–349
3. Zhu, X., Zhao, X., Burkholder, W.F., Gragerov, A., Ogata, C.M., Gottesman, M.E. and Hendrickson, W.A. (1996) Science **272**, 1606–1614
4. Wang, H., Kurochkin, A.V., Pang, Y., Hu, W., Flynn, G.C. and Zuiderweg, E.R.P. (1998) Biochemistry **37**, 7929–7940
5. Morshauser, R.C., Hu, W., Wang, H., Pang, Y., Flynn, G.C. and Zuiderweg, E.R.P. (1999) J. Mol. Biol. **289**, 1387–1403
6. Pellecchia, M., Montgomery, D.L., Stevens, S.Y., Vander Kooi, C.W., Feng, H.-P., Gierasch, L.M. and Zuiderweg, E.R.P. (2000) Nat. Struct. Biol. **7**, 298–303
7. Flaherty, K.M., Deluca-Flaherty, C. and McKay, D.B. (1990) Nature (London) **346**, 623–628
8. Harrison, C.J., Hayer-Hartl, M., Di Liberto, M., Hartl, F.-U. and Kuriyan, J. (1997) Science **276**, 431–435
9. Shi, L., Kataoka, M. and Fink, A.L. (1996) Biochemistry **35**, 3297–3308
10. Wilbanks, S.M., Chen, L., Tsuruta, H., Hodgson, K.O. and McKay, D.B. (1995) Biochemistry **34**, 12095–12106
11. Buchberger, A., Theyssen, H., Schröder, H., McCarty, J.S., Virgallita, G., Milkereit, P., Reinstein, J. and Bukau, B. (1995) J. Biol. Chem. **270**, 16903–16910
12. Montgomery, D.L., Morimoto, R.I. and Gierasch, L.M. (1999) J. Mol. Biol. **286**, 915–932
13. Burkholder, W.F., Zhao, X., Zhu, X., Hendrickson, W.A., Gragerov, A. and Gottesman, M.E. (1996) Proc. Natl. Acad. Sci. U.S.A. **93**, 10632–10637
14. Voisine, C., Craig, E.A., Zufall, N., von Ahsen, O., Pfanner, N. and Voos, W. (1999) Cell **97**, 565–574
15. Cavanagh, J., Fairbrother, W.J., Palmer, III, A.G. and Skelton, N.J. (1996) Protein NMR Spectroscopy: Principles and Practice, Academic Press, San Diego
16. Bartels, C., Xia, T.-H., Billeter, M., Güntert, P. and Wüthrich, K. (1995) J. Biomol. NMR **5**, 1–10
17. Wüthrich, K. (1986) NMR of Proteins and Nucleic Acids, Wiley, New York
18. Sehorn, M. and Witt, S.N. (2000) Biophys. J. **78**, 36A
19. Mammen, M., Choi, S.-K. and Whitesides, G.M. (1998) Angew. Chem. Int. Ed. **37**, 2754–2794

20. Booth, I.R. (1985) Microbiol. Rev. **49**, 359–378
21. Heyde, M. and Portalier, R. (1990) FEMS Microbiol. Lett. **69**, 19–26
22. Koradi, R., Billeter, M. and Wüthrich, K. (1996) J. Mol. Graph. **14**, 51–55

Biochem. Soc. Symp. **68**, 83–93
(Printed in Great Britain)

6

Validation of protein-unfolding transition states identified in molecular dynamics simulations

Valerie Daggett

Department of Medicinal Chemistry, University of Washington, Seattle, WA 98195-7610, U.S.A.

Abstract

Experimental and simulation studies can complement each other nicely in the area of protein folding. Experiment reports on the average properties of a large ensemble (approx. 10^{17}–10^{19} molecules), typically over time. Molecular dynamics simulations, on the other hand, provide detailed information for a single molecule, a component of the ensemble. By combining these approaches we can obtain not only a more complete picture of folding, but we can also take advantage of the strengths of different methods. For example, experiment cannot provide molecular structures. Molecular dynamics simulations can provide such information, but the simulations are meaningless without a linked experiment. Thus, the interrelated nature of simulation in assessing experimental assumptions and in providing structures to augment energetic descriptions, and experiment in judging whether the simulations are reasonable, provides more confidence in the resulting information about folding. This combination yields tested and testable molecular models of states that evade characterization by conventional methods. Therefore, we have explored the combined use of these methods to map folding/unfolding pathways at atomic resolution, in collaboration with Alan Fersht. Here we focus on chymotrypsin inhibitor 2, a small single-domain, two-state folding protein.

The combination of experiment and simulation is providing an unprecedented view of protein folding/unfolding. To really understand the protein folding process, all species in the pathway must be characterized, including the transient, and somewhat nebulous, transition state. Until recently, folding studies have focused primarily on intermediate states in order to piece together pathways. However, since it has been demonstrated that proteins can fold in a two-state manner without the formation of stable intermediates [1], more attention has been directed to the only 'state' between the native (N) and denatured (D) states. The protein engineering method (Φ-value analysis) is the only experimental technique currently available for probing the transition state in depth, and by combining it with molecular dynamics (MD) simulations, we can obtain a heretofore-unattainable view of these transient protein folding/unfolding states.

As with experiment, studying transition states presents problems for simulation studies. The transition state for folding/unfolding is an ensemble of high free-energy structures. Unfortunately, even if a reasonable unfolding pathway can be simulated with MD, the calculation of free energies for such a complicated process is not possible. Instead we rely on structural properties, which is the advantage of force field methods, to identify the transition state in a simulation. Using this approach, structural attributes of the transition state ensemble can be precisely delineated; however, there is no guarantee that the ensemble identified is the state of highest free energy. Since the transition state is kinetically and thermodynamically unstable, we expect the structure of the protein to change rapidly once it passes the major transition state. Therefore, we define the major transition state in simulations as the ensemble of structures immediately prior to the onset of a large structural change. The easiest way to find the transition state regions is through conformational cluster analysis [2,3]. Because we cannot determine free energies in an unrestrained MD simulation, the above definition of the transition state is neither precise nor rigorous, making extensive comparison with experiment imperative. We have attempted to challenge, and hopefully validate, our simulations in a variety of ways. The simulations were performed as predictions and the agreement with experiment has been quite satisfying; however, most comparisons are indirect, so in more recent work we have explored a number different approaches to test our transition-state assignments and, therefore, unfolding pathways, using chymotrypsin inhibitor 2 (CI2) as our model system.

CI2 contains 64 residues, comprising a single α-helix, and a mixed parallel and antiparallel β-sheet (Figure 1). CI2 was the first protein shown to fold in a two-state manner [1] and it is a good model for studying elementary folding events. As a two-state folding protein, characterization of the transition state (TS) is crucial, since it represents the only observable species between the native and denatured end points:

$$D \xrightarrow{TS} N$$

Furthermore, owing to its lack of subdomains and relatively uniform structure, with essentially the entire chain contributing to a single module of structure, CI2 represents a basic folding unit, or foldon [4]. Both unfolding and refolding have been investigated under a variety of conditions and the results are independent of direction, indicating that the transition states for folding and unfolding are identical.

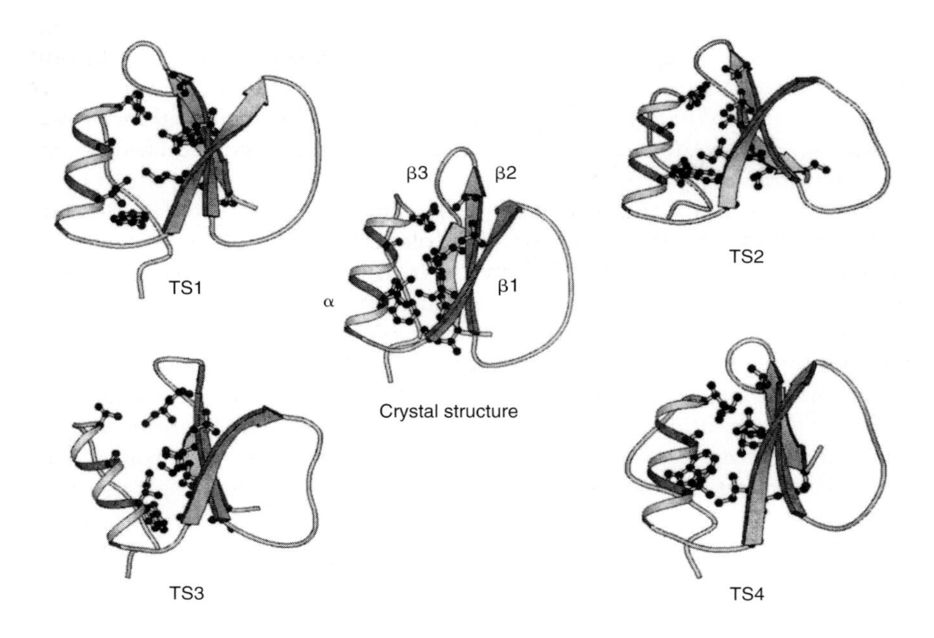

Figure 1 Secondary structure and packing interactions in the transition state models of CI2. Molscript [14] was used to prepare this figure.

The structure of the transition state of folding/unfolding has been studied by a variety of techniques, including a Φ-value analysis [5] using >100 mutations spanning the length of the protein [6,7]. Φ-values are calculated from the following equation:

$$\Phi = (\Delta G_{TS\rightarrow D} - \Delta G'_{TS\rightarrow D})/(\Delta G_{N\rightarrow D} - \Delta G'_{N\rightarrow D}) = \Delta\Delta G_{TS\rightarrow D}/\Delta\Delta G_{N\rightarrow D}$$

where $\Delta G_{TS\rightarrow D}$ and $\Delta G_{N\rightarrow D}$ are the free energies of the transition state and the native state respectively relative to the denatured state for the wild-type protein, and the corresponding terms for the mutant are indicated by a prime. $\Delta\Delta G_{N\rightarrow D}$ and $\Delta\Delta G_{TS\rightarrow D}$, therefore, are the destabilization energies of the native and other state of interest respectively caused by mutation. Consider a case where, in the transition state of unfolding, the structure of the protein at the site of mutation is the same as in the native state. In this case, the protein is immune to the effect of the mutation until after the major transition state, and the transition state is destabilized by exactly the same amount as the native state, that is $\Delta\Delta G_{TS\rightarrow D} = \Delta\Delta G_{N\rightarrow D}$ and $\Phi = 1$. Conversely, a Φ-value of 0 means that the structure of the transition state at the site of mutation is the same as in the denatured state. Intermediate values represent structures that are partially unfolded in the transition state (or a mixture of folded and unfolded).

There are a number of assumptions associated with this approach. The first is that structure is determined by energetics. While we know that structure and energy must be linked, they are not linearly related and we do not know the precise mathematical relationship between the two, which is mostly a problem in interpreting partial Φ-values. Another assumption is that the mutation merely monitors what occurs in the wild-type transition state. As such, the mutation should not change the pathway of folding/unfolding or introduce new interactions. Therefore, the choice of mutations is critical. In general, non-disruptive, conservative changes are desirable. Nonetheless, the fundamental approach appears to be sound, based on the consistency of the results of multiple mutations both at and around a site.

The Φ-values for CI2 tend to fall between 0.2 and 0.5 [6,7]. Values for residues in the α-helix, particularly at the N-terminus, and residues in the β-sheet that dock to the helix, are higher. Since the Φ-values are fractional and low, for the most part, one could imagine that the transition state is only partially structured or that there are parallel pathways of folding, and therefore parallel transition states, some with high and others with low degrees of structure. The latter situation does not exist for CI2, as determined by examining the effect of mutations on the transition state [8]. Therefore, the data suggest that the transition state of folding/unfolding of CI2 is a narrow and single ensemble of structures.

To obtain molecular models for the transition state of folding/unfolding of CI2, four unfolding simulations of CI2 (termed MD1–MD4, beginning with the crystal structure and different NMR structures) were performed at 498 K, and a transition state was identified from each (Figure 1). The simulations were performed, in parallel with the experimental studies, in a blind manner, that is, they were performed as predictions. The transition states identified in the four independent simulations (TS1–TS4) are similar overall and the unfolding pathways only diverge after the transition state. The transition states have the following characteristics: the hydrophobic core is considerably weakened; the secondary structure, particularly the β-sheet, is frayed; and packing of the secondary structure is disrupted considerably (Figure 1). The overall structure of the transition state is closer to the native than to the unfolded state.

To determine the position of the transition state during unfolding, structures before and after the identified transition state position were quenched at lower temperature and their conformational properties were monitored. For the first simulation, MD1, the movement from the first cluster occurs at ~225 ps. Structures before (220 ps) and after (230 ps) this position were simulated at a lower temperature, 335 K. If the choice of transition state is correct, then the structure preceding the transition state should become more native-like. Since the 230 ps structure should occur after the transition state, it should encounter a significant free-energy barrier to refolding, and it is expected to demonstrate less native-like behaviour than the 220 ps structure. The results agree with the expectations. During the 3 ns of the simulation beginning with the 220 ps structure, the C_α-atom root mean square deviation (RMSD) drops from 5 to 3 Å and the solvent-accessible surface area decreases by 800 $Å^2$ (Figure 2). In contrast, the RMSD remains high in the simulation beginning with the 230 ps structure and the protein does not collapse (Figure 2).

For comparison with the experimental Φ-values, 'computer-mutations' were made to the transition state structures and the difference in packing con-

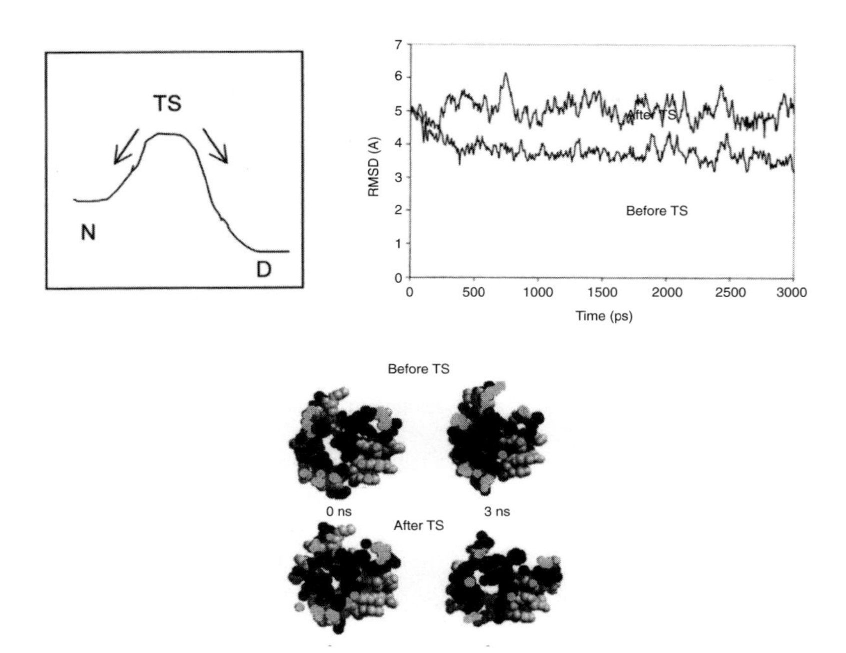

Figure 2 Behaviour of structures before and after the putative transition state (TS1) when simulated under quasi-native conditions. Non-polar residues are shown in black in the structures on the right. UCSF MidasPlus [15] was used to prepare this figure

tacts between the wild-type and mutant proteins in the transition and native states were evaluated to determine a Φ_{MD}-value [2,3]. This approach provides good agreement with experiments for the 11 hydrophobic deletion mutants described by Otzen and co-workers [6]: the correlation coefficient for comparison of Φ_{MD} with the experimental Φ-values is 0.88 for the most thoroughly studied transition state model, TS1. In fact, the best agreement with experiment is obtained when the four transition states are pooled ($R = 0.94$), highlighting the fact that the transition state is an ensemble of related structures [3]. In more recent studies, we evaluated the structure of the wild-type transition states, which allows us to test the assumption that the mutations are merely probes of what is occurring in the wild-type protein. The structure of the wild-type transition state can be evaluated semi-quantitatively via local structure indices, termed *S* values. For each residue, an *S*-value is calculated that is a product of the local percentage of tertiary and secondary structure in the transition state structures relative to the native state [9]. The calculated *S*-values agree with the experimental values (Figure 3) for the entire protein, giving a correlation coefficient of 0.89. The agreement lends support to the assumption that the protein engineering approach need not change dramatically the folding process and can report on the behaviour of the wild-type protein. We note also that independent unfolding simulations of CI2 performed by [10], using a different force field, are consistent with the results described here.

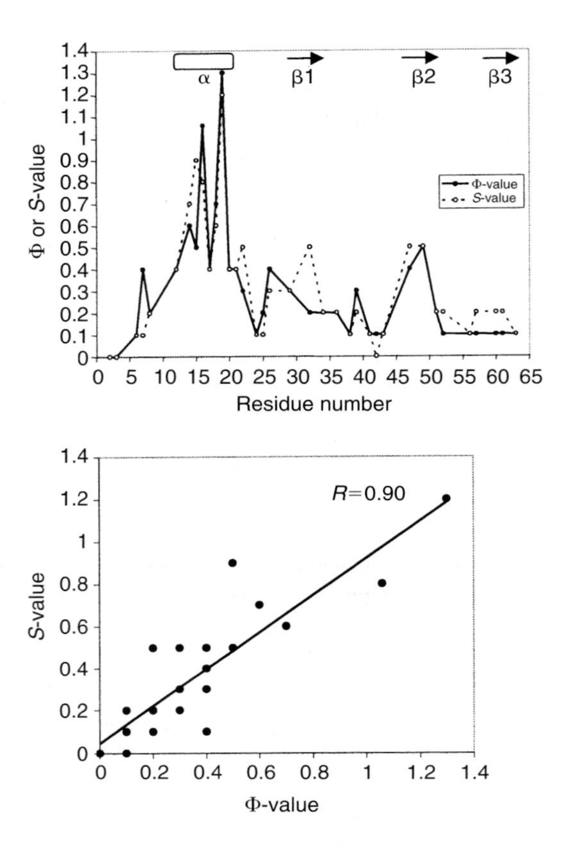

Figure 3 Comparison of experimental Φ-values and the average calculated *S*-values for the CI2 transition states.

While the MD-generated models are in very good agreement with the experimental Φ-values, we still desire an approach to calculate directly free energies for the conformational states sampled during MD. To this end, we have undertaken free-energy-perturbation calculations [11] of the hydrophobic deletion mutants investigated by Otzen and co-workers [6]. For a direct calculation of Φ-values from the MD simulation, we made the mutations in the native state, in an MD-generated transition-state model, and in different denatured conformations after the protein fully unfolds. The mutations are made in each of the conformational states, and the corresponding free energy for the mutation is calculated. Through use of a thermodynamic cycle, as shown in Figure 4, these calculated free energies (the vertical lines) are then related to the experimental values (the horizontal lines). This is the first time such calculations have been performed on structures along an unfolding pathway. In addition, the availability of these detailed models of the denatured state allow us to test the validity of the more common approach of using tripeptides to model the unfolded state in calculations of protein stability. The calculated free-energy changes for each of the eight different mutations for the native

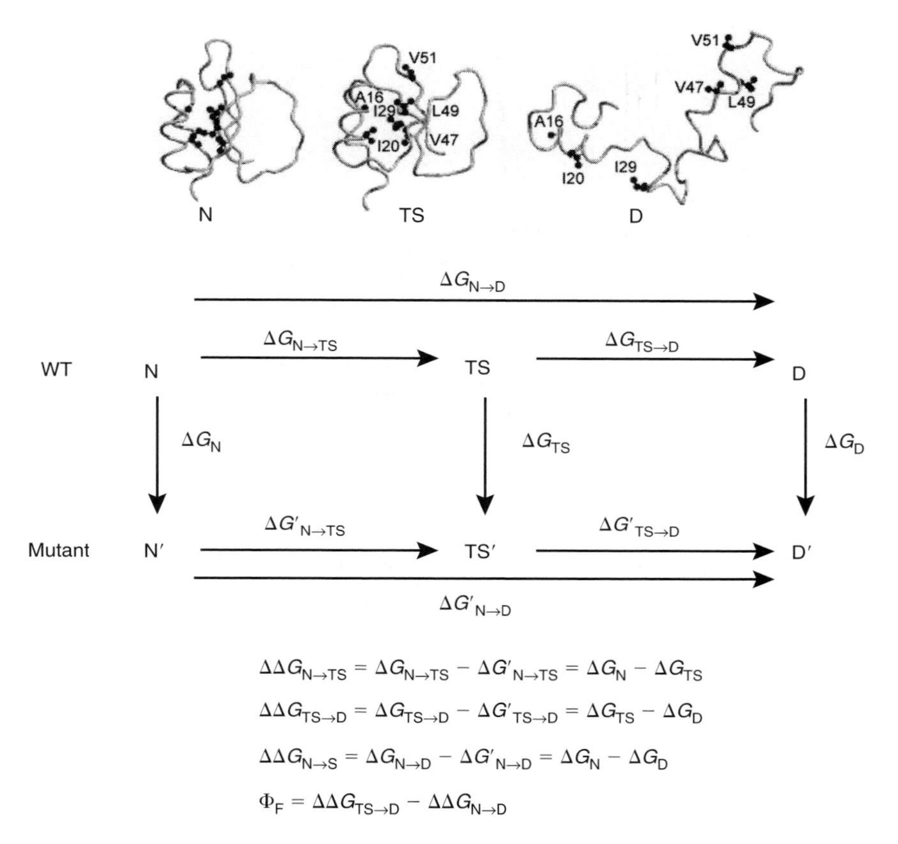

Figure 4 Free-energy cycle used to relate results for making hydrophobic deletion mutants in the core of CI2. The wild-type (WT) transition state and denatured state structures shown were taken from the MD1 simulation [2,3]. UCSF MidasPlus [15] was used to prepare this figure.

(ΔG_N), transition (ΔG_{TS}) and denatured (ΔG_D) states are listed in Table 1, along with the calculated destabilization energies ($\Delta\Delta G_{N\rightarrow D}$, $\Delta\Delta G_{N\rightarrow TS}$ and $\Delta\Delta G_{TS\rightarrow D}$) and the Φ-values. Both the $\Delta\Delta G$- and Φ-values agree very well with the experimental results, yielding a correlation coefficient of 0.85 for comparison of the experimental and calculated Φ-values [11].

The agreement between our calculations and the experimental results for the A16G and V19A mutations is worth special attention. A16G is the only mutation to yield an experimental Φ-value of 1. Our calculations are in excellent agreement with this (Table 1). The V19A mutation results in a negative Φ-value, which suggests that this side-chain makes more contacts in the transition state than the native state. Other mutations that delete one methylene group (I20V, I29V) are also in good agreement with experiment (Table 1). On the other hand, there is less agreement when the mutations involve more than two methylene groups. In the case of V47A and L49A, the calculated free-energy changes are

Table 1 Calculated (Calc.) and experimental (Expt) free-energy changes (kcal/mol) for hydrophobic deletion mutations in CI2

Mutant	ΔG_N	ΔG_D	ΔG_{TS}	$\Delta\Delta G_{N\to D}$		$\Delta\Delta G_{N\to TS}$		$\Delta\Delta G_{TS\to D}$		Φ_F	
				Calc.	Expt	Calc.	Expt	Calc.	Expt	Calc.	Expt
A16G	2.28	1.10	2.17	1.18	1.09	0.11	−0.20	1.07	1.15	0.91	1.06
V19A	1.68	1.09	1.03	0.59	0.49	0.77	0.86	−0.06	−0.13	−0.10	−0.26
I20V	2.64	1.42	1.71	1.22	1.30	0.93	0.57	0.29	0.52	0.24	0.40
I29V	2.45	1.04	1.75	1.41	1.11	0.70	0.69	0.71	0.19	0.50	0.37
V47A	2.60	0.74	1.85	1.86	4.93	0.73	3.81	1.13	1.02	0.61	0.21
L49A	2.64	0.63	1.78	2.01	3.80	0.86	1.45	1.15	2.11	0.57	0.53
V51A	4.24	1.92	2.77	2.32	1.98	1.47	1.00	0.85	0.49	0.37	0.25
I57A	5.42	1.80	2.60	3.62	4.29	2.82	3.46	0.80	0.36	0.22	0.08

The ΔG_{TS} values are averages for separate calculations of TS structures from three independent unfolding simulations. The G_D values are averages over five different denatured-state structures. See [11] for details. 1 kcal = 4.184 kJ.

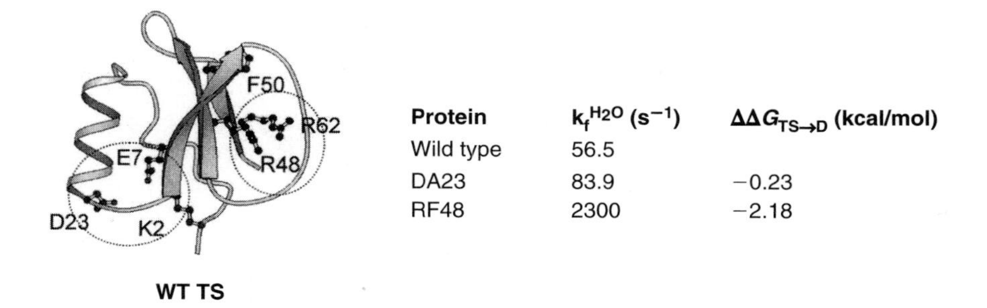

Protein	$k_f^{H_2O}$ (s^{-1})	$\Delta\Delta G_{TS\rightarrow D}$ (kcal/mol)
Wild type	56.5	
DA23	83.9	−0.23
RF48	2300	−2.18

Figure 5 Unfavourable electrostatic interactions in the transition state of CI2 (a negative charge at the C-terminus of the helix and charge repulsion in the active site loop) and the mutations made to relieve strain in the transition state. The resulting experimental values are given to the right [13]. Molscript [14] was used to prepare this figure. WT TS, wild-type transition state. 1 kcal = 4.184 kJ.

consistently lower than the experiment results. There are slight changes in packing upon creation of these larger voids in both the native and transition-state structures, such that the Φ-values remain in good agreement.

Despite the good agreement between simulation and experiment for the transition state of folding/unfolding of CI2, the high temperature used in the simulations and the limited sampling warrant concern. For example, Oliveberg et al. [12] have shown that the transition state of CI2 becomes more native-like at high temperatures. Therefore, further experimental tests of the simulations are desirable. To this end, the simulated structures of the transition state were used to design faster-folding CI2 mutants. The models pinpoint a number of unfavourable local interactions at the C-terminus of the α-helix, and in the protease-binding-loop region of CI2. Therefore, the prediction is that if unfavourable interactions are removed via mutation, folding will proceed more rapidly.

The first region investigated was the C-terminus of the helix. Asp-23 stabilizes the native protein by forming a salt bridge with Lys-2. However, the presence of an aspartate residue at this position in an isolated helix has a destabilizing effect because of unfavourable interactions with the carbonyl groups at the end of the helix — the helix macrodipole. The simulated transition state has essentially an isolated helix when this salt bridge is broken (Figure 5). We predicted that a DA23 mutant should fold faster than wild-type CI2 through stabilization of the transition state for folding (Figure 5). This is indeed the case; the refolding rate constant increases from 56 s^{-1} for wild type, to 84 s^{-1} for DA23 [13] (Figure 5). These increases are especially significant considering that overall destabilization of CI2 generally leads to a decrease in the rate constant for folding [6].

The second region investigated contains a cluster of positively charged residues, Arg-43, Arg-46, Arg-48, and Arg-62 (Figure 5). The hydrophobic side chains of these residues stack together, bringing their guanidinium groups into close proximity to one other and thereby causing electrostatic strain. This strain is partly relieved by a network of hydrogen bonds with the carboxylate group of the C-terminal residue, Gly-64, in the native state. This loop region is expanded and more loosely packed in the transition state (Figures 1 and 5). All four transition state models show three or four of the Arg residues in proximity, and the native salt bridges and favourable ionic interactions are not well formed in the transition state. An imperfectly formed salt bridge in a non-polar environment is unstable, which places severe constraints on transition states, and hence on the rate of protein folding. The removal of some of the unfavourable electrostatic interactions between the positively charged guanidinium groups, and improvement of non-polar packing in the region, would, therefore, be expected to stabilize the transition state. An RF48 mutation was made and the rate of folding increased from 56 to 2300 s^{-1} to yield the fastest-folding form of CI2 thus far. The mutations described above (and others) were designed to yield faster-folding versions of CI2 based solely on the MD-generated transition-state models. By removing unfavourable interactions identified in the models, the rate of folding increased up to 40-fold.

Conclusions

High-temperature MD simulations have been used to map protein unfolding pathways at the atomic level. Transition-state ensembles have been identified from these simulations. While these ensembles are in good, even quantitative, agreement with experiment, they are identified based on structural features, not energetics. Thus, fortuitous agreement is possible. A number of tests of these transition-state ensembles are described here. Structures before and after the putative transition state have been simulated at lower temperatures. The structure prior to the transition state collapses and becomes more native-like with the temperature quench. In contrast, the structure taken after the putative transition state experiences a barrier to refolding and does not collapse or become more native-like. Free-energy perturbation calculations have been performed recently using transition- and denatured-state structures from the simulations to allow direct comparison with the experimental free-energy changes upon mutation — both stability and kinetic measurements. The calculated and experimental free-energy changes, as well as the Φ-values, are in excellent agreement. Finally, unfavourable interactions have been identified in the transition state models. These interactions have been removed by mutation to yield faster-folding versions of CI2. This completes the cycle by first performing the simulations and experiments separately, by comparing the results and showing the complementary nature of the two approaches, by performing further computational tests of the models, and finally by making predictions based on the model structures and testing those predictions in the laboratory.

Partial support of this work was provided by the NIH (GM 50789 to V.D.).

References

1. Jackson, S.E. and Fersht, A.R. (1991) Biochemistry **30**, 10428–10435
2. Li, A. and Daggett, V. (1994) Proc. Natl. Acad. Sci. U.S.A. **91**, 10430–10434
3. Li, A. and Daggett, V. (1996) J. Mol. Biol. **257**, 412–429
4. Panchenko, A.R., Luthey-Schulten, A. and Wolynes, P.G. (1996) Proc. Natl. Acad. Sci. U.S.A. **93**, 2008
5. Matouschek, A., Kellis, Jr, J.T., Serrano, L. and Fersht, A.L. (1989) Nature (London) **340**, 122–126
6. Itzhaki, L.S., Otzen, D.E. and Fersht, A.R. (1995) J. Mol. Biol. **254**, 260–288
7. Otzen, D.E., Itzhaki, L.S., El Masry, N.F., Jackson, S.E. and Fersht, A.R. (1994) Proc. Natl. Acad. Sci. U.S.A. **91**, 10422–10425
8. Matouschek, A., Otzen, D.E., Itzhaki, L.S., Jackson, S.E. and Fersht, A.R. (1995) Biochemistry **34**, 13656–13662
9. Daggett, V., Li, A., Itzhaki, L.S., Otzen, D.E. and Fersht, A.R. (1996) J. Mol. Biol. **257**, 430–440
10. Lazaridis, T. and Karplus, M. (1997) Science **278**, 1928–1931
11. Pan, Y.P and Daggett, V. (2001) Biochemistry **40**, 2723–2731
12. Oliveberg, M., Tan, Y.-J., Silow, M. and Fersht, A.R. (1998) J. Mol. Biol. **277**, 933–943
13. Ladurner, A.G., Itzhaki, L.S., Daggett, V. and Fersht, A.R. (1998) Proc. Natl. Acad. Sci. U.S.A. **95**, 8473–8478
14. Kraulis, P.J. (1991) J. Appl. Crystallogr. **24**, 946–950
15. Ferrin, T.E., Huang, C.C., Jarvis, L.E. and Langridge, R. (1988) J. Mol. Graphics **6**, 13–27

Biochem. Soc. Symp. 68, 95–110
(Printed in Great Britain)

7

Structure, stability and folding of the α-helix

Andrew J. Doig*[1], Charles D. Andrew*, Duncan A.E. Cochran*, Eleri Hughes*, Simon Penel*, Jia Ke Sun*, Benjamin J. Stapley*, David T. Clarke† and Gareth R. Jones†

*Department of Biomolecular Sciences, University of Manchester Institute of Science and Technology, P.O. Box 88, Manchester M60 1QD, U.K., and †Daresbury Laboratory, Daresbury, Warrington, Cheshire WA4 4AD, U.K.

Abstract

Pauling first described the α-helix nearly 50 years ago, yet new features of its structure continue to be discovered, using peptide model systems, site-directed mutagenesis, advances in theory, the expansion of the Protein Data Bank and new experimental techniques. Helical peptides in solution form a vast number of structures, including fully helical, fully coiled and partly helical. To interpret peptide results quantitatively it is essential to use a helix/coil model that includes the stabilities of all these conformations. Our models now include terms for helix interiors, capping, side-chain interactions, N-termini and 3_{10}-helices. The first three amino acids in a helix (N1, N2 and N3) and the preceding N-cap are unique, as their amide NH groups do not participate in backbone hydrogen bonding. We surveyed their structures in proteins and measured their amino acid preferences. The results are predominantly rationalized by hydrogen bonding to the free NH groups. Stabilizing side-chain–side-chain energies, including hydrophobic interactions, hydrogen bonding and polar/non-polar interactions, were measured accurately in helical peptides. Helices in proteins show a preference for having approximately an integral number of turns so that their N- and C-caps lie on the same side. There are also strong periodic trends in the likelihood of terminating a helix with a Schellman or α_L C-cap motif. The kinetics of α-helix folding have been studied with stopped-flow deep ultraviolet circular dichroism using synchrotron radiation as the light source; this gives a far superior signal-to-noise ratio than a conventional instrument. We find that poly(Glu), poly(Lys) and alanine-based

[1]To whom correspondence should be addressed.

peptides fold in milliseconds, with longer peptides showing a transient overshoot in helix content.

Structure of the α-helix

Regular local folds in polypeptide chains form the secondary structure of proteins. The α-helix was first described by Pauling et al. in 1950 [1], and their model was quickly supported by X-ray analysis of haemoglobin [2]. Irrefutable proof of the existence of the α-helix was provided by the protein crystal structure of myoglobin [3], in which all secondary structure is helical. α-Helices were found subsequently in nearly all globular proteins. It is the most abundant secondary structure, with approx. 30% of residues found in α-helices [4].

A helix combines a linear translation with an orthogonal circular rotation. In the α-helix the linear translation is a rise of 5.4 Å per turn of the helix and the circular rotation is 3.6 residues per turn. Side chains spaced at $i,i+3$, $i,i+4$ and $i,i+7$ are, therefore, close in space, and interactions between them can affect helix stability. Spacings of $i,i+2$, $i,i+5$ and $i,i+6$ place the side-chain pairs on opposite faces of the helix, thereby, avoiding any interaction. The helix is stabilized primarily by $i,i+4$ hydrogen bonds between backbone amide groups.

The conformation of a polypeptide can be described by the backbone dihedral angles φ and ψ. Most φ/ψ combinations are sterically excluded, leaving only the broad β-region and narrower α-region. One reason why the α-helix is so stable is that a succession of the sterically allowed α φ- and ψ-angles naturally position the backbone NH and CO groups towards each other for hydrogen-bond formation. It is possible that a succession of the most stable conformation of an isolated residue could point hydrogen-bond donors towards other donors, making secondary structure unfavourable. One reason why polypeptides may have been selected as the polymer of choice for building functional molecules is that the sterically most stable conformations also give strong hydrogen bonds.

The residues at the N-terminus of the α-helix are designated as N′-N-cap-N1-N2-N3-N4 etc., where the N-cap is the residue with non-helical φ/ψ angles immediately preceding the N-terminus of an α-helix and N1 is the first residue with helical φ/ψ angles [5]. The C-terminal residues are similarly designated C4-C3-C2-C1-C-cap-C′ etc. The N1, N2, N3, C1, C2 and C3 residues are unique because their amide groups participate in $i,i+4$ backbone–backbone hydrogen bonding on one side only. The need for these groups to form hydrogen bonds has powerful effects on helix structure and stability.

Methods

Since Davies [6] first showed that some amino acids occur more often than others in the helix, work on the properties of the α-helix has included extracting structural and statistical information from Protein Data Bank (PDB) files. A more detailed examination of amino acid preferences in the α-helix revealed a greater complexity, because the frequency of occurrence of an amino acid depends upon its position in the helix. Argos and Palau [7], and Richardson and Richardson [5], showed that the helix has significantly different residue frequencies for the cap, N-terminal and C-terminal and helix interior positions. The continuing rapid growth of the PDB allows more statistically significant information to be extracted. For example, our studies on hydrogen bonding at 3_{10}-helix N-caps [8] would have been meaningless if there were only 20 non-homologous, high-resolution crystal structures available.

An attractive way to study experimentally the thermodynamics and kinetics of the helix/coil equilibrium is to use a peptide that forms monomeric helices in aqueous solution. In the 1960s, it was believed that a peptide would need to have hundreds of residues to form an appreciable amount of helix. Polymers, containing only Lys or Glu, with heterogeneous lengths were therefore used as models to study helix formation.

In 1971, Brown and Klee [9] showed that the S-peptide of ribonuclease formed an α-helix despite having fewer than 20 amino acids, although this system was not used as a model to understand the helix until Baldwin's work in the late 1980s. In 1989, Marqusee et al. [10] showed that a peptide composed predominantly of Ala, with Lys at every fifth position to confer solubility, could form a monomeric α-helix in water. Peptides based on this sequence have been of major importance in elucidating the factors that affect helix stability [11–13]. At the same time, the development of site-directed mutagenesis techniques has enabled helices in proteins to be mutated directly (for example [14–18]).

Helix/coil theory

Peptides that form helices in solution do not show a simple, two-state equilibrium between a fully folded and a fully unfolded structure. Instead, they form a complex mixture of all helix, all coil or, most frequently, central helices with frayed-coil ends. In order to interpret experimental data on helical peptides and make theoretical predictions on helices it is essential to use a helix/coil theory that considers every possible conformation of the peptide and includes every possible interaction in the peptide that can affect the stability of any conformation.

Our helix/coil theory is based on the Lifson–Roig model [19]. Each residue is assigned a conformation of helix (h) or coil (c), depending on whether it has helical φ/ψ angles. Every conformation of a peptide of N residues can, therefore, be written as a sequence of N c or h. Residues are assigned statistical weights depending on their conformations and the conformations of surrounding residues. A residue in an h-conformation with an h on either side has a weight of w. This can be thought of as an equilibrium constant between the helix interior and the coil. Coil residues thus have a weight of 1. In order to form an $i,i+4$ hydrogen bond in a helix, three successive residues need to be fixed in a helical conformation. M consecutive helical residues will therefore have $M - 2$ hydrogen bonds. The two residues at the helix termini (i.e. those in the centre of chh or hhc conformations) are, therefore, assigned weights of v. The ratio w:v gives the effect of hydrogen bonding {1.7:0.036 for Ala [20] or

$-RT \ln(1.7/0.036) = -2.1$ kcal·mol^{-1} (8.8 kJ·mol^{-1})}(R, the Gas Constant; T, temperature). A helical segment with M residues has a weight of v^2w^{M-2} and a population in the equilibrium of v^2w^{M-2} divided by the sum of the weights of every conformation (i.e. the partition function). In this way, the population of every conformation is calculated and all properties of the helix/coil equilibrium, e.g. helical population of each residue, mean number of hydrogen bonds formed, probability that each residue is at the N-cap, are evaluated. A two-state analysis of the helix/coil equilibrium, in particular, gives serious errors. All our experiments on helical peptides are interpreted with helix/coil models to gain information on properties, such as capping and side-chain interaction energies.

We have greatly extended the Lifson–Roig model to include new factors that affect helix stability, by assigning weights to additional conformations [21–23]. The N-cap is the residue with non-helical φ/ψ angles preceding the helix. Hence, its weight is assigned to the central residue in a *cch* triplet. Similarly, C-cap weights are assigned to *hcc* triplets, $i,i+4$ side-chain interactions to *hhhhh* quintets (because five consecutive helical residues are necessary to form the interaction), $i,i+3$ side chain interactions to *hhhh* quartets, N1 preferences to *ch* conformations, N2 preferences to *chh* conformations and N3 preferences to *chhh* conformations. We have also developed 3_{10}-helix [24,25], π-helix [24] and β-sheet [26] models by classifying residues into 3_{10}-, π- or β-conformations instead of h. These models are essential if studies of the helix/coil equilibrium are to be placed on a quantitative basis. The models predict accurately the helix contents of peptides, provided that all the required energetic terms are known (e.g. the control peptides in Table 1). If the peptide contains a side-chain interaction that has not been measured precisely, as will be the case for most natural sequences, the AGADIR algorithm [27] can be used instead, since this includes an estimate of every side-chain interaction energy.

Helical peptides, coiled coils and protein helices have been used extensively to determine the intrinsic preferences of the amino acids to be in interior helical positions, giving satisfying agreement [28]. Extensive comparisons of peptide experiments with those on proteins indicate that the results of peptide-based work are directly applicable to proteins. In particular, measurement of helix propensities in the same helix of a peptide and in a protein gave excellent agreement [29,30].

N-cap structures and energies

Preferences for the N-cap position were found by measuring the helix contents of peptides with sequence NH_2-X-Ala-Lys-Ala-Ala-Ala-Ala-Lys-Ala-Ala-Ala-Ala-Lys-Ala-Ala-Gly-Tyr-$CONH_2$ (X represents any amino acid) [31]. Within the helix/coil equilibrium, the N-terminal amino acid will be in a coil or at the N-cap. By applying helix/coil theory [21] to the peptide sequence, and with mean helix content determined by CD, we determined the N-cap energies of each amino acid (Table 2). The results correlated well with statistical preferences in crystal structures [5], and with results from site-directed mutagenesis of N-caps [14,16], again showing that results from

Table 1 Peptides used to measure side chain interaction energies

Name	Sequence	Experimental helix content	Predicted helix content (no side chain interaction)	Side chain interaction energy (kcal·mol^{-1})
Phe–Met	Ac-Tyr-Gly-Ala-Ala-Lys-Ala-Ala-Phe-Ala-Lys-Ala-Met-Ala-Ala-Lys-Ala-Ala-NH_2	37	22	−0.8
Phe–Met control	Ac-Tyr-Gly-Ala-Ala-Lys-Ala-Phe-Ala-Ala-Lys-Ala-Met-Ala-Ala-Lys-Ala-Ala-NH_2	22	23	–
Met–Phe	Ac-Tyr-Gly-Ala-Ala-Lys-Ala-Ala-Met-Ala-Lys-Ala-Phe-Ala-Ala-Lys-Ala-Ala-NH_2	32	22	−0.5
Met–Phe control	Ac-Tyr-Gly-Ala-Ala-Lys-Ala-Met-Ala-Ala-Lys-Ala-Phe-Ala-Ala-Lys-Ala-Ala-NH_2	19	22	–
Gln–Asn	Ac-Ala-Gln-Ala-Ala-Ala-Ala-Ala-Gln-Ala-Ala-Ala-Asn-Ala-Ala-Ala-Ala-Gln-Gly-Tyr-NH_2	55	39	−0.5
Gln–Asn control	Ac-Ala-Gln-Ala-Ala-Ala-Ala-Gln-Ala-Ala-Ala-Ala-Asn-Ala-Ala-Ala-Ala-Gln-Gly-Tyr-NH_2	37	36	–
Asn–Gln	Ac-Ala-Gln-Ala-Ala-Ala-Ala-Ala-Asn-Ala-Ala-Ala-Gln-Ala-Ala-Ala-Ala-Gln-Gly-Tyr-NH_2	37	39	−0.1
Asn–Gln control	Ac-AlaGlnAlaAlaAlaAlaAsnAlaAlaAlaAlaGln-Ala-Ala-Ala-Ala-Gln-Gly-Tyr-NH_2	33	36	–
Ile–Lys	Ac-Ala-Lys-Ala-Ile-Ala-Ala-Ala-Lys-Ala-Ile-Ala-Ala-Ala-Lys-Ala-Lys-Ala-Gly-Tyr-NH_2	51	37	−0.25
Ile–Lys control	Ac-Ala-Lys-Ile-Ala-Ala-Ala-Ala-Lys-Ile-Ala-Ala-Ala-Ala-Lys-Ala-Lys-Ala-Gly-Tyr-NH_2	37	39	–

Interacting side chains are underlined. Ac, acetyl; 1 kcal = 4.184 kJ.

peptides are directly relevant to proteins. It is remarkable that Asn and Gln are at opposite extremes with respect to N-cap preference, despite both having the same amide functional group in their side chains. This suggests that an Asn to Gln substitution at an N-cap would be a very non-conservative change and that treating chemically similar amino acids as if they have identical properties can be misleading. The terms 'helix breaker' or 'helix maker' for the amino acids are not useful. A substitution could increase the stability of a helix when in the middle of a sequence, while being destabilizing at the N-terminus (e.g. Ala to Asp).

Our empirical results were rationalized by surveying N-caps in crystal structures [8]. We found very strong rotamer preferences that are unique to N-cap sites. The following rules are generally observed for N-capping in α-helices: (1) Thr and Ser N-cap side chains adopt the g^- rotamer, hydrogen bond to the N3 NH group and have ψ restricted to $164 \pm 8°$; (2) Asp and Asn N-cap side chains either adopt the g^- rotamer and hydrogen bond to the N3 NH group with $\psi = 172 \pm 10°$, or adopt the t rotamer and hydrogen bond to both the N2 and N3 NH groups with $\psi = 107 \pm 19°$. With all other N-caps, the side chain is found in the g^+ rotamer so that the side chain does not interact unfavourably with the N-terminus by blocking solvation and ψ is unrestricted. An $i,i+3$ 3_{10}-helix hydrogen bond from N3 NH to the N-cap backbone C=O is more likely to form at the N-terminus when an unfavourable N-cap is present.

N1N2N3 structures and energies

We found strong structural preferences that are unique to the helix N1, N2 and N3 positions, by examining statistical preferences, rotamers, hydrogen bonding and solvent accessibilities [32]. The 'good N2' amino acids Gln, Glu, Asp, Asn, Ser, Thr and His preferentially form i,i or $i,i+1$ hydrogen bonds to the backbone, although this effect is reduced when 'good N-caps' (Asp, Asn, Ser, Thr and Cys) that compete for these hydrogen bond donors are present. We found a number of specific side-chain–side-chain interactions between N1 and N2 or between the N-cap and N2 or N3, e.g. between Arg(N-cap) and Asp-(N2). The strong energetic and structural preferences found for N1, N2 and N3, which differ greatly from positions within helix interiors, suggest that these sites should be treated individually in any consideration of helical structure in peptides or proteins. The dominant interaction that is common to all our studies of helix N-termini is the formation of hydrogen bonds to the otherwise unpaired NH groups of N1, N2 and N3, as foreseen by Presta and Rose in 1988 [33]. This appears to be a much stronger effect than the helix dipole, which would imply that Asp and Glu are most favoured at N-terminal locations.

After extending helix/coil theory to include N1, N2 and N3 [23], we calculated the N1 preferences of the amino acids [23], using peptides with sequence Ac-X-Ala_4-Gln-Ala_4-Gln-Ala-Ala-Gly-Tyr-$CONH_2$ (Table 1). We found that Ala, Asp and Glu have the highest preferences for the N1 position and that positively charged groups are disfavoured, in agreement with the helix dipole model. Most amino acids all have similar preferences, 0.5 $kcal \cdot mol^{-1}$ (2.1 $kJ \cdot mol^{-1}$) less than Ala. Gln, Asn and Ser, therefore, do not stabilize the helix

Table 2 Position-dependent N-terminal preferences

Residue	ΔG for transfer from coil to N-cap (kcal·mol^{-1}) [31]	ΔG for transfer from coil to N1 (kcal·mol^{-1})
Asn	−1.7	Too high to measure
Asp$^-$	−1.6	0.7
Acetyl	−1.4	
Cys$^-$	−1.4	
Trp	−0.6	1.1
Gly	−1.2	1.7
Ser	−1.2	1.1
Tyr	−1.5	Too high to measure
Thr	−0.7	1.2
His^0	−0.7	1.4
Leu	−0.7	1.1
Phe	−0.9	2.1
Glu$^-$	−0.7	0.8
Ile	−0.5	1.2
Pro	−0.4	1.3
Met	−0.3	1.2
Ala	0.0	0.7
Arg$^+$	−0.1	1.4
Val	−0.1	1.3
Lys$^+$	0.1	1.4
Gln	2.5	1.2

ΔG; free energy values.

when they are at N1, despite frequent hydrogen bonding to backbone NH groups, probably because these hydrogen bonds are non-linear and, therefore, weaker [32]. In contrast, hydrogen bonds formed by N-cap side chains are close to linear and so are stronger [8].

Side-chain interactions in helices

Ala-based peptides offer a useful system to accurately measure interaction energies between side chains on the surface of a solvent-exposed helix. We compared control peptides, for which all the helix/coil parameters are known, with an isomeric peptide with the side chains of interest spaced $i,i+4$ (Table 1). For peptides close to 50% helicity, the mean helix content, measured by CD, is very sensitive to a small change in stability; therefore, the interaction energy can be measured accurately. The helix contents of the control peptides are predicted successfully, using helix/coil theory, to within experimental error (±3%), while the only unknown factor in the test peptide is the energy of the interaction. Adjusting the side-chain interaction energy until the calculated helix content agrees with experiment allows one to determine the energy. Table 1 shows our measurements for hydrophobic interactions (Phe–Met and

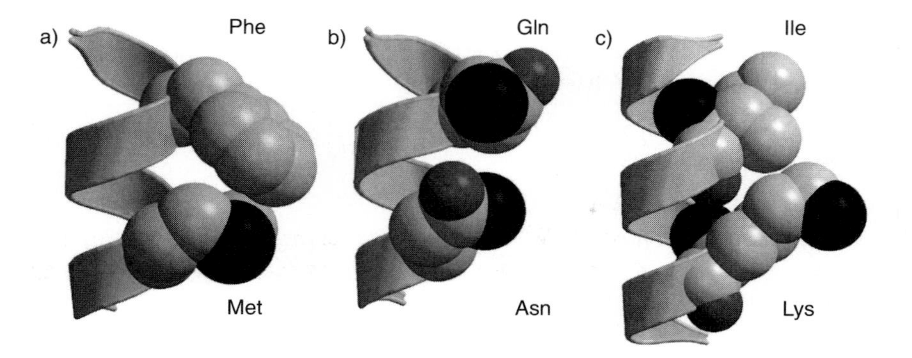

Figure 1 Structures of *i,i*+4 side-chain interactions in the α-helix. All side chains are shown in their most populated rotamers when in helices in protein crystal structures. (a) Phe–Met (hydrophobic). Phe-262 and Met-266 from 1clc.pdb. (b) Gln–Asn (hydrogen bonding). Gln-127 and Asn-131 from 1ako.pdb. (c) Ile–Lys (non-polar/polar). Ile-121 and Lys-125 from 153l.pdb.

Met–Phe), hydrogen bonds (Gln–Asn) and polar/non-polar interactions (Ile–Lys) (Figure 1). The side-chain order in the helix can be critical. The Gln–Asn energy is very different from that of Asn–Gln, as the geometric requirements for hydrogen bond formation are difficult to fulfil in the Asn–Gln orientation [34]. The Gln–Asn interaction involves a hydrogen bond between amide groups and, therefore, shows the strength of the backbone amide hydrogen bond, without doubt the most important hydrogen bond in proteins. Despite the flexible side chains being on the surface of a helix exposed to water, the interaction has a substantially stabilizing effect, indicating that the hydrogen bond makes a considerable contribution to protein stability. It is, perhaps, surprising that the Ile–Lys interactions are stabilizing, as bonds between hydrophobic and charged side chains would be expected to be unfavourable. Examination of these interactions in crystal structures, however, shows that the Cγ2 carbon of Ile and the Cϵ carbon of the Lys interact.

Side chain interactions in helices studied by other groups include Asp–Lys, Glu–His, Glu–Lys, Phe–His, His–Glu, Lys–Asp, Lys–Glu, Lys–Lys, Lys–Tyr, Gln–Asp, Trp–His, Tyr–Leu and Tyr–Val. Fernandez-Recio and Sancho provide a useful list of their energies [35]. All have a stabilizing effect, apart from the Lys–Lys repulsion.

Rotamer strain energy in α-helices and a new view of protein stability

Protein folding is opposed by loss of conformational entropy when side chains change from occupying multiple rotamers in the unfolded state, to a single rotamer when folded. A number of independent techniques have shown that that the mean conformational free-energy change on folding ($T\Delta S$) is

1 kcal·mol^{-1} (4.184 kJ·mol^{-1}) per side chain [36]. The conformational free-energy change on folding the backbone is of a similar magnitude [37].

The energetic changes from restricting side chain torsional motion are more complex than simply loss of conformational entropy, however. A second force that opposes protein folding arises when a side chain in the folded state is not in its lowest energy rotamer, giving rotameric strain. χ-Strain energy results from a dihedral angle being shifted from the most stable location within a rotamer when a protein folds. We calculated the energy of a side chain as a function of its dihedral angles in a poly(Ala) helix using CHARMm [38]. Using these energy profiles, we quantified conformational entropy, rotameric strain energy and χ-strain energy for all 17 amino acids with side chains in α-helices. We can calculate these values for any amino acid in a helix interior, as a function of its side chain dihedral angles. The algorithm is implemented on a web site (http://wolf.bi.umist.ac.uk/helix/create_pdb.html). The mean rotameric strain energy is 0.7 kcal·mol^{-1} (2.93 kJ·mol^{-1}) per residue and the mean χ-strain energy is 0.6 kcal·mol^{-1} (2.51 kJ·mol^{-1}) per residue. Loss of conformational entropy opposes folding by a mean of 1.1 kcal·mol^{-1} (4.60 kJ·mol^{-1})per residue [36], so the mean total force opposing restricting a side chain into a helix is 2.4 kcal·mol^{-1} (10.04 kJ·mol^{-1}).

It is widely believed that the main force opposing protein folding is loss of conformational entropy. By quantifying strain energy, however, we demonstrate that strain energy introduced when a side chain is restricted away from its most stable conformation is at least as large as conformational entropy. We have restricted our work to interior positions in α-helices, but have no reason to expect other sites to be significantly different. Strain from distorting bond angles and lengths may also be important. Therefore, conformational entropy estimates alone greatly underestimate the forces opposing protein folding.

The traditional view of protein stability assumes that proteins are stabilized by the hydrophobic effect and destabilized by loss of conformational entropy, while other possible energetic factors, such as hydrogen bonding, are negligible. We suggest that this 'old view of protein stability' is incompatible with the latest experimental data. A wealth of empirical evidence, reviewed by Myers and Pace [39], has demonstrated that hydrogen bonding does indeed stabilize proteins. Similarly, one way in which side chains can interact favourably is by hydrogen bonding (Table 1), and hydrogen bonding is the dominant force at the helix N-terminus. Our work indicates that the introduction of strain when a protein folds should not be neglected. We propose a 'new view of protein stability', whereby proteins are stabilized by the hydrophobic effect and hydrogen bonding, and destabilized by strain and loss of conformational entropy. All four terms are of comparable magnitude.

Periodicity in α-helix lengths and C-cap preferences

We surveyed protein crystal structures for α-helix lengths (Figure 2) [40]. While the data did not fit to any simple statistical distribution (not surprisingly, as selecting helix lengths is essentially a problem of packing cylinders in

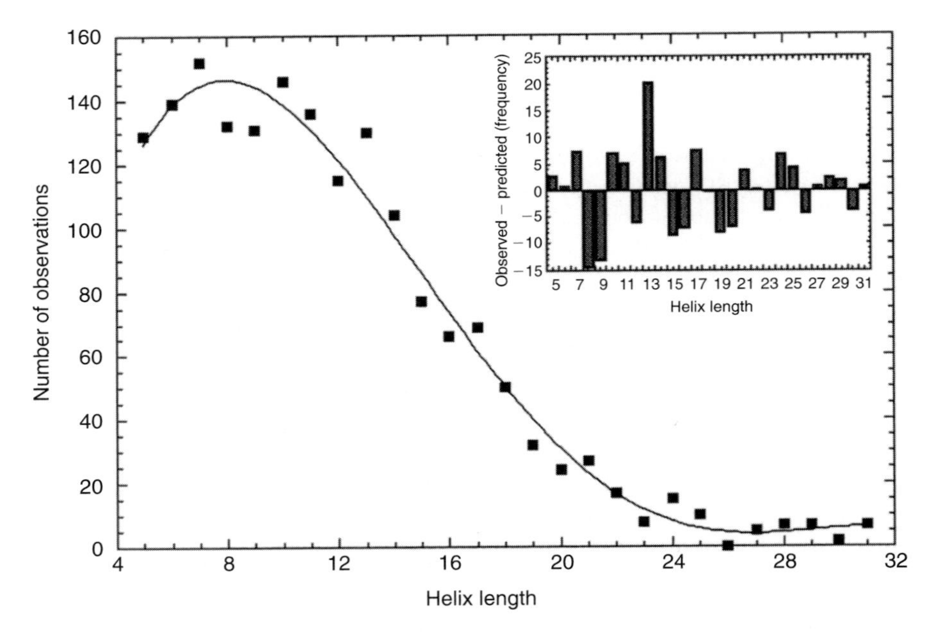

Figure 2 Frequency distribution of helix lengths. The number of observations of helices is plotted as a function of helix length. Data are fitted to the equation $y = -51.58 + 60.419x - 5.9201x^2 + 0.20116x^3 - 0.0022942x^4$. The inset shows the residuals from this smoothing function to reveal the 3.6-residue periodicity. Adapted from Penel, S., et al. (1999) *J. Mol. Biol.* **293**, 1211–1219, with permission from Academic Press.

spheres), we observed that helices show a preference to have close to an integral number of turns. With an integral number of turns, the helix will have its N- and C-caps on the same side, and so can be placed on the protein surface.

Helices can be usefully subdivided into either 'favoured length' with 6, 7, 10, 11, 13, 14, 17, 18, 21, 22, 24, 25, 28, 29 or 31 residues, or 'disfavoured length' with 8, 9, 12, 15, 16, 19, 20, 23, 26, 27 or 30 residues. This is reflected at the C-terminus, where favoured-length helices that are often on the protein surface (with the C-cap buried) prefer non-polar side chains at C4 and polar amino acids at C2, while disfavoured-length helices prefer non-polar amino acids at C2.

Helix C-termini often form C-capping motifs with unusual conformations to maximize hydrogen bonding to free CO groups [41]. We found strong periodic trends in the likelihood of terminating a helix with a Schellman (Figure 3) or α_L C-capping motif. These can be rationalized by the preference for a non-polar side chain at C3 in these motifs, which favours the placing of C3 on the buried side of the helix. We suggest that algorithms designed to predict helices or C-capping in proteins should include a weight for the helix length.

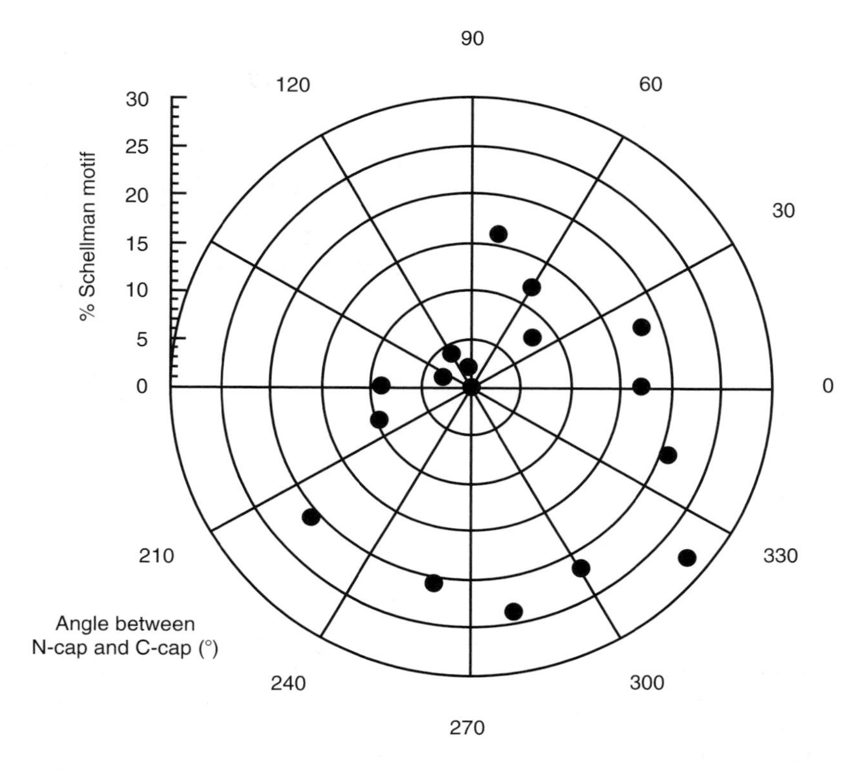

Figure 3 Polar co-ordinates plot of percentage of helices that terminate with a Schellman motif versus angle between the N-cap and C-cap (calculated as 100° per residue). The Schellman motif is overwhelmingly preferred when the angle between the N- and C-caps is 40–220°. Reproduced from Penel, S., et al. (1999) *J. Mol. Biol.* **293**, 1211–1219, with permission from Academic Press.

Folding kinetics of the α-helix

There are two microscopic rates in helix folding [42]: first, the fast propagation of an existing helix by the addition of a single residue to the end of a helix; and secondly, the rate of initiation of a new helix, presumably by the formation of a single turn, stabilized by one $i,i+4$ hydrogen bond. This will be slow, since it requires the entropically unlikely event of the simultaneous restriction of three successive residues, as we have observed with helix/coil theory (see above). Relaxation times for the helix/coil transition of Glu and Lys homopolymers have been measured by temperature-jump experiments [43–45]. Temperature-jump infra-red spectroscopy [46] and N-terminal reporter group fluorescence studies [47] have also been applied to measure the kinetics of unfolding of a 21-residue poly(Ala)-based helical peptide. These results have been used to derive a rate for extension of helices by a single turn of 1×10^7–7×10^{10} s^{-1} and to infer very high initiation rates for the coil-to-helix transition.

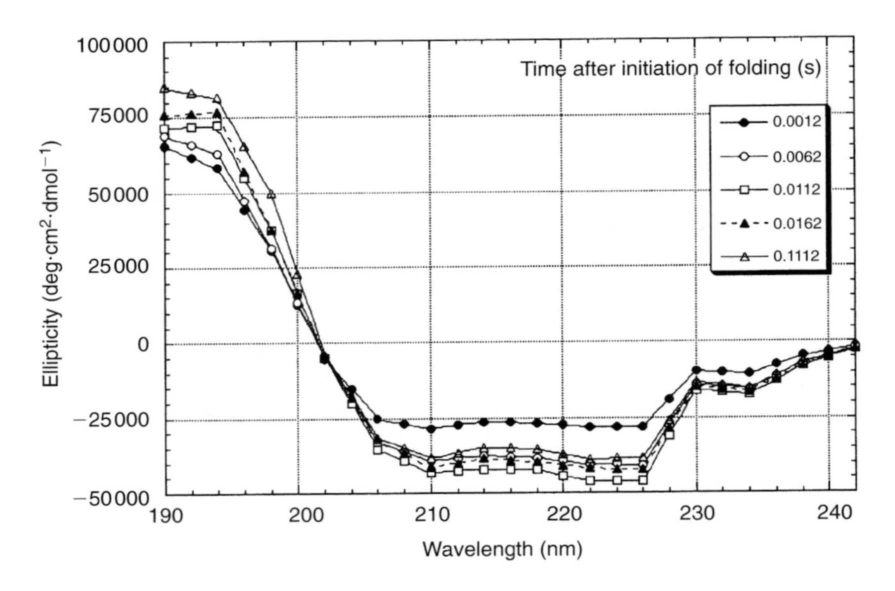

Figure 4 CD spectra of 4.4 kDa poly(Glu) during folding at 293 K. Spectra were acquired every 2.5 ms for 110 ms, although only five spectra are shown here for clarity. There is an isodichroic point at 203 nm, indicating a two-state transition between coil and helix. The kinetics differ above and below the isodichroic point, with the higher wavelength data showing an overshoot, and the lower wavelength data none. We have no explanation for this. Reproduced from Clarke, D.T., et al. (1999) *Proc. Natl. Acad. Sci. U.S.A.* **96**, 7232–7237, with permission from Academic Press.

We measured directly the rate of nucleation of α-helices from the denatured state [48]. Helix formation in AK16 (sequence Ac-Tyr-Gly-Ala-Ala-Lys-Ala-Ala-Ala-Ala-Lys-Ala-Ala-Ala-Ala-Lys-Ala-NH_2) was initiated by a 10-fold dilution with 5 M guanidinium chloride (GuHCl). We also studied poly(L-Lys) and poly(L-Glu), which form α-helices when neutral and random coils when charged. Initiation of helix folding was therefore performed by a pH jump, from 8.0 to 11.5 for poly(Lys) and from 8.0 to 3.3 for poly(Glu). Helix content was monitored by stopped-flow CD spectroscopy using synchrotron radiation. The use of synchrotron radiation instead of a UV lamp improves the signal-to-noise ratio by more than 100-fold in the 190 nm range. We were, therefore, able to acquire CD spectra of 4.4 kDa poly(Glu) from 190 to 242 nm every 2.5 ms with a resolution of 2 nm as it folded (Figure 4). This result is unachievable with a conventional instrument. Kinetic data were acquired for all the peptides at 222 or 226 nm, where there is a strong negative signal in the α-helix CD spectrum. The AK16 data fitted a single exponential with no overshoot (Figure 5). The folding rate and final helix content (θ_{222} at equilibrium) both decrease as temperature increases.

We studied the rate of unfolding of 4.4k Da poly(Glu) by following a pH jump, and of AK16 by jumping from 0.00 to 5.45 M GuHCl, and found that the

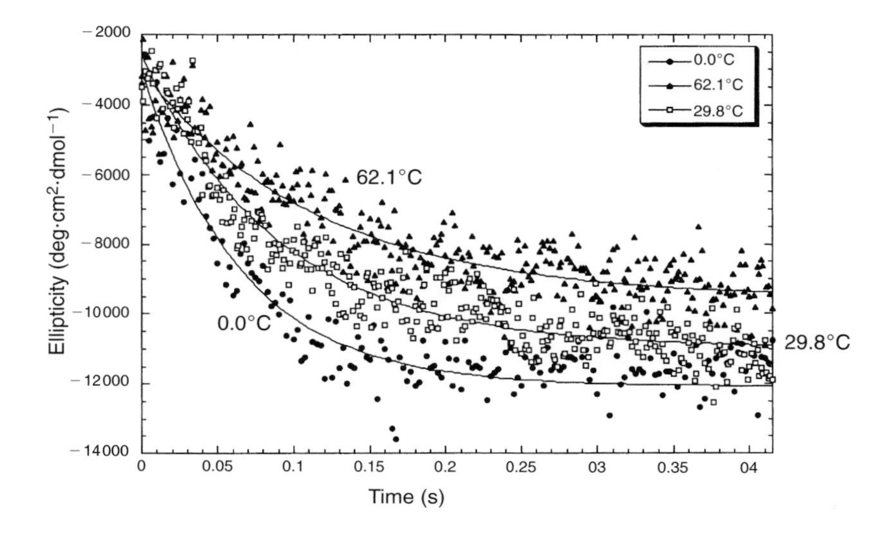

Figure 5 Refolding kinetics of α-helix formation in the AK16 peptide (Ac-Tyr-Gly-Ala-Ala-Lys-Ala-Ala-Ala-Ala-Lys-Ala-Ala-Ala-Ala-Lys-Ala-NH_2) monitored by stopped-flow CD at 222 nm. Data were fitted to the equation $\theta = A(1-e^{-kt})+\theta(U)$. *A*, a constant; *k*, rate constant; *U*, an unfolded state. Reproduced from Clarke, D.T., et al. (1999) *Proc. Natl. Acad. Sci. U.S.A.* **96**, 7232–7237, with permission from Academic Press.

peptide was completely unfolded within the dead time of the stop-flow device, in agreement with the earlier work. In contrast, our helix folding rates are far lower. This may be because the rate-determining step in our kinetics is the initiation of a new helix, while previous workers studied a shift in the helix/coil equilibrium within highly helical peptides, where the observed rate amplitude arises largely from crossing the propagation barrier. Alternatively, helix initiation rates could be very sequence-dependent. Comparison with our rates suggests that helix propagation is at least 10^5 times faster than initiation. Helix nucleation is surprisingly slow; perhaps the rate-limiting step is the formation of a larger structure than simply one helix turn with one hydrogen bond.

The poly(Lys) and poly(Glu) data fit well to a three-state mechanism (coil→intermediate→equilibrium). Data in the 206 to 234 nm range all gave similar results. The ellipticity at 226 nm (θ_{226}) in poly(Glu), poly(Lys) and AQ28 transiently achieves a minimum that is considerably greater than that of the native state after 2–10 ms. There appears to be a large and rapid generation of high ellipticity at 226 nm followed by relaxation to equilibrium with a smaller ellipticity. Aggregation is ruled out by invariance of kinetics with a 100-fold variation in concentration. Transient 3_{10}-helix or poly(Pro) II formation are not seen in the CD spectra (Figure 4) that have an isodichroic point. It appears most likely that the results show the slow gain and subsequent loss of monomeric α-helical structure.

We suggest that the overshoot arises from two factors: (1) while helix initiation is slow, the conversion of an existing helix into two helices will also be slow because this requires the simultaneous breaking of three backbone hydrogen bonds; and (2) the mean helix content of a peptide consisting of a single helix is higher than the mean helix content of the peptide with several helices present at equilibrium, as shown by consideration of the effects of the 'one-sequence approximation' within helix/coil theory [49–51]. The fully charged peptide is entirely unfolded and extended. After initiation, the helix will propagate rapidly, forming a very long helix that includes nearly all the residues. This single-helix conformation will have a high helix content, giving the maximum ellipticity observed in the kinetics. At equilibrium the peptide will have many conformations containing two or more helices, perhaps in helix–turn–helix or anti-parallel coiled-coil structures. The transition from the single helix conformation to multiple helices requires either the nucleation of a new helix in a coil region or the breakage of the long helix. Both of these steps are slow; therefore, the rate from the peak of the overshoot to equilibrium is low and there is a decrease in helix content. There is no overshoot for the peptide AK16, as conformations with multiple helices are insignificantly populated because of its short length. Hence, these kinetics simply show the relaxation rate for crossing the nucleation barrier.

It is not clear whether the rates we find will be similar in other helix sequences. Poly(Lys) and poly(Glu) are poly(Pro) II helix when non-helical [52] and folding from this structure may be anomalously slow. None of these peptides have stabilizing side-chain interactions that could help initiate helix formation, unlike protein helix sequences. Repulsion between the Lys side chains in AK16 may also hinder transition-state formation. Further work is needed to test these ideas.

Conclusion

Experimental studies on Ala-based peptides, combined with surveys of protein crystal structures, have provided many quantitative insights into the structure, stability and folding of this most important secondary structure. We now have the intrinsic propensities of the amino acids for helix interiors, capping sites and N1. The measurement of side-chain interaction energies, combined with the propensity data, allows the accurate prediction of all properties of the helix/coil equilibrium using helix/coil theory. Helices recorded in the PDB help to rationalize the peptide results by revealing statistically preferred structures. Furthermore, the PDB yields new structural properties not present in peptides, such as a periodic preference for length. We propose a 'new view of protein stability' whereby proteins are stabilized by the hydrophobic effect and hydrogen bonding, and destabilized by loss of conformational entropy and importantly by strain. Stopped-flow CD studies show that some peptide helices fold surprisingly slowly; it remains to be seen how general this observation is, and thus how relevant it is to proteins.

We thank the Biotechnology and Biological Sciences Research Council, Wellcome Trust, the Central Laboratories of the Research Councils, Royal Society, Nuffield Foundation, and Merck Sharp & Dohme for funding.

References

1. Pauling, L., Corey, R.B. and Branson, H.R. (1951) Proc. Natl. Acad. Sci. U.S.A. **37**, 205–211
2. Perutz, M.F. (1951) Nature (London) **167**, 1053–1054
3. Kendrew, J.C., Dickerson, R.E., Strandberg, B.E., Hart, R.G., Davies, D.R., Phillips, D.C. and Shore, V.C. (1960) Nature (London) **185**, 422–427
4. Barlow, D.J. and Thornton, J.M. (1988) J. Mol. Biol. **201**, 601–619
5. Richardson, J.S. and Richardson, D.C. (1988) Science **240**, 1648–1652
6. Davies, D.R. (1964) J. Mol. Biol. **9**, 605–609
7. Argos, P. and Palau, J. (1982) Int. J.Pept. Protein Res. **19**, 380–393
8. Doig, A.J., MacArthur, M.W., Stapley, B.J. and Thornton, J.M. (1997) Protein Sci. **6**, 147–155
9. Brown, J.E. and Klee, W.A. (1971) Biochemistry **10**, 470–476
10. Marqusee, S., Robbins, V.H. and Baldwin, R.L. (1989) Proc. Natl. Acad. Sci. U.S.A. **86**, 5286–5290
11. Scholtz, J.M. and Baldwin, R.L. (1992) Annu. Rev. Biophys. Biomol. Struct. **21**, 95–118
12. Baldwin, R.L. (1995) Biophys. Chem. **55**, 127–135
13. Rohl, C.A. and Baldwin, R.L. (1998) Methods Enzymol. **295**, 1–26
14. Bell, J.A., Becktel, W.J., Sauer, U., Baase, W.A. and Matthews, B.W. (1992) Biochemistry **31**, 3590–3596
15. Serrano, L., Neira, J.L., Sancho, J. and Fersht, A.R. (1992) Nature (London) **356**, 453–455
16. Serrano, L., Sancho, J., Hirshberg, M. and Fersht, A.R. (1992) J. Mol. Biol. **227**, 544–559
17. Horovitz, A., Matthews, J.M. and Fersht, A.R. (1992) J. Mol. Biol. **227**, 560–568
18. Blaber, M., Zhang, X.-J. and Matthews, B.W. (1993) Science **260**, 1637–1640
19. Lifson, S. and Roig, A. (1961) J. Chem. Phys. **34**, 1963–1974
20. Rohl, C.A., Chakrabartty, A. and Baldwin, R.L. (1996) Protein Sci. **5**, 2623–2637
21. Doig, A.J., Chakrabartty, A., Klingler, T.M. and Baldwin, R.L. (1994) Biochemistry **33**, 3396–3403
22. Stapley, B.J., Rohl, C.A. and Doig, A.J. (1995) Protein Sci. **4**, 2383–2391
23. Sun, J.K., Penel, S. and Doig, A.J. (2000) Protein Sci. **9**, 750–754
24. Rohl, C.A. and Doig, A.J. (1996) Protein Sci. **5**, 1687–1696
25. Sun, J.K. and Doig, A.J. (1998) Protein Sci. **7**, 2374–2383
26. Sun, J.K. and Doig, A.J. (2000) J. Phys. Chem. B, **104**, 1826–1836
27. Muñoz, V. and Serrano, L. (1994) Nat. Struct. Biol. **1**, 399–409
28. Pace, C.N. and Scholtz, J.M. (1998) Biophys. J. **75**, 422–427
29. Myers, J.K., Pace, C.N. and Scholtz, J.M. (1997) Proc. Natl. Acad. Sci. U.S.A. **94**, 2833–2837
30. Regan, L. (1997) Proc. Natl. Acad. Sci. U.S.A. **94**, 2796–2797
31. Doig, A.J. and Baldwin, R.L. (1995) Protein Sci. **4**, 1325–1336
32. Penel, S., Hughes, E. and Doig, A.J. (1999) J. Mol. Biol. **287**, 127–143
33. Presta, L.G. and Rose, G.D. (1988) Science **240**, 1632–1641
34. Stapley, B.J. and Doig, A.J. (1997) J. Mol. Biol. **272**, 465–473
35. Fernandez-Recio, J. and Sancho, J. (1998) FEBS Lett. **429**, 99–103
36. Doig, A.J. and Sternberg, M.J. (1995) Protein Sci. **4**, 2247–2251
37. D'Aquino, J.A., Gómez, J., Hilser, V.J., Lee, K.H., Amzel, L.M. and Freire, E. (1996) Proteins **25**, 143–156
38. Brooks, B.R., Bruccoleri, R.E., Olafson, B.D., States, D.J., Swaminathan, S. and Karplus, M. (1983) J. Comp. Chem. **4**, 187–217

39. Myers, J.K. and Pace, C.N. (1996) Biophys. J. **71**, 2033–2039
40. Penel, S., Morrison, R.G., Mortishire-Smith, R.J. and Doig, A.J. (1999) J. Mol. Biol. **293**, 1211–1219
41. Aurora, R. and Rose, G.D. (1998) Protein Sci. **7**, 21–38
42. Schwarz, G. (1965) J. Mol. Biol. **11**, 64–77
43. Bosterling, B. and Engel, J. (1979) Biophys. Chem. **9**, 201–209
44. Hamori, E. and Scheraga, H.A. (1967) J. Phys. Chem. **71**, 4145–4150
45. Lumry, R., Legare, R. and Miller, W.G. (1964) Biopolymers **2**, 489–498
46. Williams, S., Causgrove, T.P., Gilmanshin, R., Fang, K.S., Callendar, R.H., Woodruff, W.H. and Dyer, R.B. (1996) Biochemistry **35**, 691–697
47. Thompson, P.A., Eaton, W.A. and Hofrichter, J. (1997) Biochemistry **36**, 9200–9210
48. Clarke, D.T., Doig, A.J., Stapley, B.J. and Jones, G.R. (1999) Proc. Natl. Acad. Sci. U.S.A. **96**, 7232–7237
49. Schellman, J.A. (1958) J. Phys. Chem. **62**, 1485–1494
50. Qian, H. and Schellman, J.A. (1992) J. Phys. Chem. **96**, 3987–3994
51. Muñoz, V. and Serrano, L. (1997) Biopolymers **41**, 495–509
52. Woody, R.W. (1992) Adv. Biophys. Chem. **2**, 37–79

Biochem. Soc. Symp. **68**, 111–123
(Printed in Great Britain)

8

Guidelines for the assembly of novel coiled-coil structures: α-sheets and α-cylinders

John Walshaw, Jennifer M. Shipway and Derek N. Woolfson[1]

Centre for Biomolecular Design and Drug Development, School of Biological Sciences, University of Sussex, Falmer BN1 9QG, U.K.

Abstract

The coiled coil is a ubiquitous motif that guides many different protein–protein interactions. The accepted hallmark of coiled coils is a seven-residue (heptad) sequence repeat. The positions of this repeat are labelled *a-b-c-d-e-f-g*, with residues at *a* and *d* tending to be hydrophobic. Such sequences form amphipathic α-helices, which assemble into helical bundles via knobs-into-holes interdigitation of residues from neighbouring helices. We wrote an algorithm, SOCKET, to identify this packing in protein structures, and used this to gather a database of coiled-coil structures from the Protein Data Bank. Surprisingly, in addition to commonly accepted structures with a single, contiguous heptad repeat, we identified sequences with multiple, offset heptad repeats. These 'new' sequence patterns help to explain oligomer-state specification in coiled coils. Here we focus on the structural consequences for sequences with two heptad repeats offset by two residues, i.e. *a/f'-b/g'-c/a'-d/b'-e/c'-f/d'-g/e'*. This sets up two hydrophobic seams on opposite sides of the helix formed. We describe how such helices may combine to bury these hydrophobic surfaces in two different ways and form two distinct structures: open 'α-sheets' and closed 'α-cylinders'. We highlight these with descriptions of natural structures and outline possibilities for protein design.

Introduction

The coiled coil is a ubiquitous protein-structure motif, which guides and cements a wide variety of protein–protein interactions [1]. Over the last decade, the leucine zipper [2,3] has provided the best experimental system for studying coiled-coil folding, structure and design. Through this work the leucine zipper

[1]To whom correspondence should be addressed.

has assumed the position of archetypal coiled coil. However, the majority of coiled coils do not assemble as short, parallel dimers like the leucine zippers (Figure 1a). Rather, they form a wide variety of assemblies in which oligomer state, helix length and helix orientation vary. Moreover, many coiled coils assemble further, and beyond the primary coiled-coil interaction.

Studies on wild type, mutant and designer leucine-zipper peptides have, nonetheless, proved pivotal in reaching our current understanding of coiled-coil folding, structure and assembly. For example, the crystal structure of the leucine-zipper region of the yeast transcriptional activator GCN4 provided the first high-resolution view of a coiled coil [4], and confirmed Crick's 40-year-old, 'knobs-into-holes' (KIH) theory for coiled-coil packing [5]. Crick's earlier work had provided inspiration for determining the seven-residue signature of coiled-coil sequences, which was first achieved for tropomyosin [6]. In turn, this prompted much of Hodges' significant coiled-coil design work [7,8]. Similarly, the crystal structure of the leucine zipper sparked studies that led to the development and testing of rules for oligomer-state selection [9–12] and partner selection in coiled coils [13–16], and also a number of successful protein designs [8,17–19].

The question is: how well does what we have learned about the relatively straightforward leucine-zipper system relate to larger, more-complicated coiled-coil assemblies? This is important because of the aforementioned structural diversity of natural coiled coils. In addition, for design studies, we would like to know if we have exhausted all topology and oligomer-state possibilities for coiled-coil assemblies. One aspect of our own research is to extend and develop theories for coiled-coil assemblies, and to apply and test these in protein design [18–20].

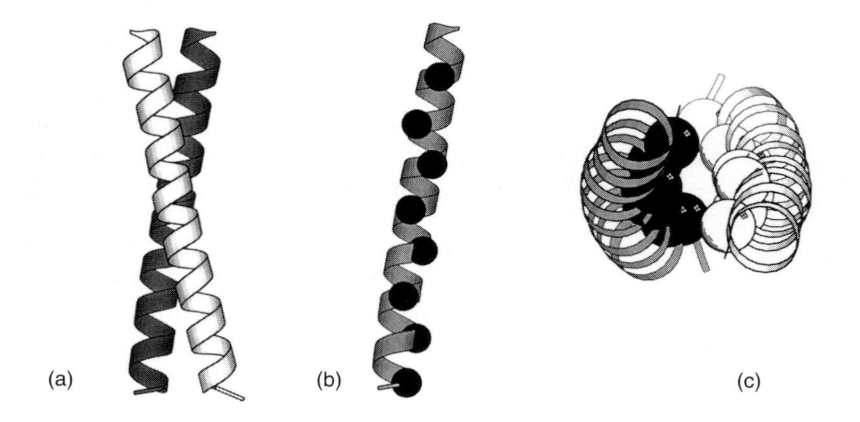

Figure 1 The GCN4 leucine-zipper. (a) Ribbon diagram showing the dimeric coiled-coil fold. (b) Highlighting the C_α atoms of the hydrophobic-core (*a* and *d*) residues for one helix. (c) The orthogonal view of the intact dimer showing the core residues. This, and subsequent, protein-structure ribbon diagrams were created using MOLSCRIPT [39].

Heptad-repeat sequences and KIH packing

Despite the variety of lengths, oligomer states and topologies observed for coiled-coil structures, their sequences usually display similar patterns known as heptad repeats [1]. The seven residues of a heptad are labelled *a-b-c-d-e-f-g* and the residues at *a* and *d* are usually hydrophobic. A contiguous array of heptads gives a heptad repeat in which hydrophobic side-chains alternate every third and fourth residue. Breaks in heptad repeats are known. These can either be irregular [21] or regular, to the extent that they alter the overall repeating pattern in the sequence [20]. Furthermore, coiled-coil-like structures without any canonical heptads have been postulated and even designed successfully [22–24]. For sequences of natural amino acids at least, non-heptad inserts appear to destabilize coiled coils (M.R. Hicks and D.N. Woolfson, unpublished work). It is possible that, particularly for larger coiled-coil proteins, non-heptad repeats modulate the stabilities of these assemblies. In addition, they may contribute also to partner selection [20]. These points aside, the heptad repeat is the most abundant, and appears to be the most appropriate, building block for the coiled coil. Indeed, it is the accepted hallmark of the coiled-coil sequences.

The structural consequences of a seven-residue repeat were first pointed out by Crick [5]. Briefly, in a heptad repeat an average alternating pattern of one hydrophobic side chain in every 3.5 residues occurs. This almost matches the 3.6 residues per turn of the α-helix. Thus, the hydrophobic residues of a coiled coil fall on the same side of the helix, to give an amphipathic structure (Figure 1b). The marriage of two such hydrophobic seams results in helical assemblies, which are stabilized by the burial of hydrocarbon from water (Figure 1c). However, because the average separation of hydrophobic residues does not match the α-helical repeat precisely, successive *a* and *d* residues fall short of their counterparts in the preceding heptad, and the hydrophobic seam winds around the surface of the helix. Consequently, in order to maintain a buried interface the helices have to wrap, or supercoil, around each other, which led to the term 'coiled coil' (Figure 1).

A further significant discovery by Crick was that the side chains in coiled-coil interfaces mesh extremely effectively (Figure 2) [5]. For parallel structures, residues at *a* and *d* form 'knobs' on one helix that dock into 'holes' on the partner, which are diamonds of residues made by $d_{-1}g_{-1}ad$ and by $adea_{+1}$, respectively (Figure 3). The subscripts, +1 and −1, refer to positions in heptads following and preceding respectively the heptad providing the knob. In the antiparallel case the situation is reversed, with *a* and *d* knobs fitting into $adea_{+1}$ and $d_{-1}g_{-1}ad$ holes. These arrangements are examples of KIH packing and differ from the looser 'ridges-into-grooves' packing associated with other helix–helix interactions, particularly those in globular proteins [25].

Since Crick made his proposals, it has been shown that, with some caveats, the interlacing of C_α positions (Figure 2) is a general feature of helix–helix packing [26], and is not restricted to the context of heptad repeats and coiled coils. However, true KIH packing involves the interlacing of side chains [27]. The nature of the interdigitating interface formed by this interlacing is quite distinct from the aforementioned ridges-into-grooves helix-packing, and

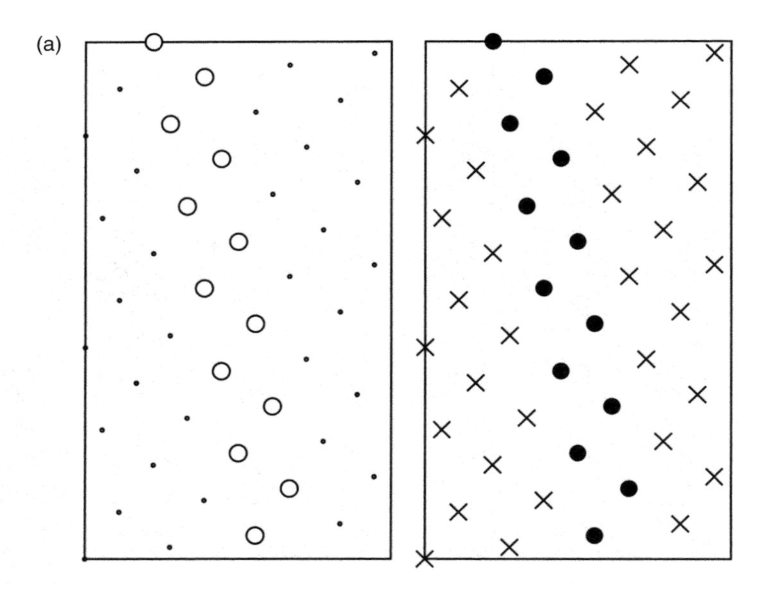

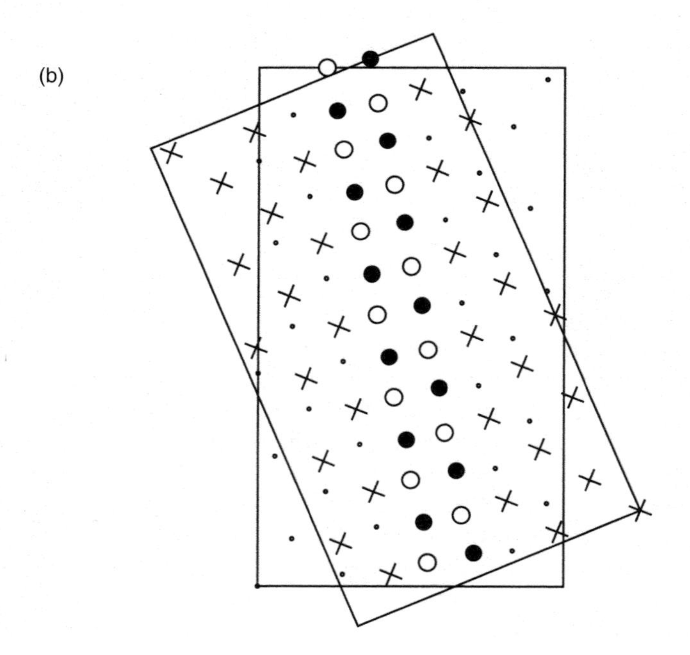

Figure 2 Helical nets illustrating the interdigitation of residues at a coiled-coil interface. (a) Nets for the individual helices. The C_α positions are marked by crosses and dots, and the *a* and *d* residues highlighted as open and filled circles. (b) Flipping the left-hand net over and inclining it at approx. $-20°$ relative to the other allows intimate interlacing of the two sets of *a* and *d* residues.

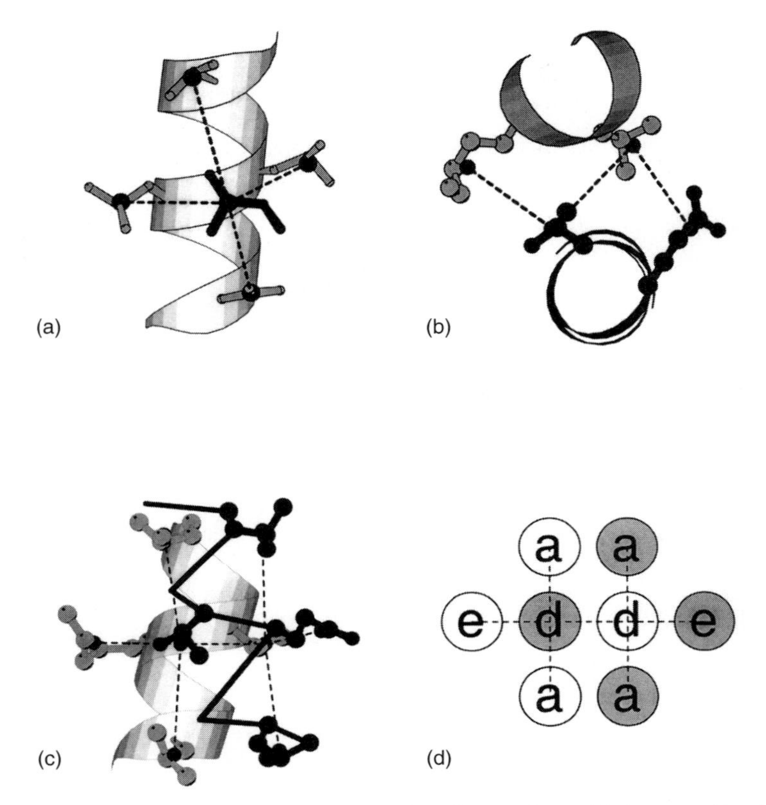

Figure 3 KIH interactions at a coiled-coil interface. (a) A single *d* knob (heavy shading) fitting into $adea_{+l}$ hole of a partnering helix. (b) Top, cut-away view of (a) showing the *dde* layer. (c) Eight side chains forming two complementary KIH interactions. (d) A schematic representation of (c).

we can distinguish the two types irrespective of the C_α positions using our own software, SOCKET (see below).

Another distinguishing feature of KIH packing is that it leads to helix–helix packing angles (Ω) of approx. 20°, which are much slighter than packing angles that predominate in globular helix–helix interactions. For heptad-based coiled coils, $\Omega \approx -20°$, which denotes left-handed supercoiling (Figures 1a and 2). This can be understood from the argument above: one heptad repeat falls short of two complete turns (7.2 residues) of a right-handed α-helix and, therefore, the resulting supercoil is in the opposite sense to twist direction of the helix. By contrast, coiled coils made from helices with pure 11-residue, hendecad, repeats have a right-handed twist and Ω is positive [20,22–24].

The slight packing-angles of coiled coils lead to KIH interdigitation of side chains over long helical distances [28,29]. This intimate packing is almost certainly responsible for the stabilities of leucine zippers and related structures, which are unusually high given their small size and the fact that they are oligomers. This point, coupled with our excellent understanding of coiled-coil folding, structure and assembly, opens up considerable opportunities for the design of self-assembling systems based on small, synthetically accessible peptides.

Sequences and structures with multiple, offset heptad repeats

We developed SOCKET to recognize KIH packing and assign unambiguously heptad registers in protein structures. We applied SOCKET to all the structures in the Protein Data Bank (PDB [30]) to collect a complete database of true coiled-coil structures. Details of SOCKET and our database will be presented elsewhere. We focus here on an interesting observation that we made during the analysis of the protein structures and the structural possibilities that this opens up for natural and designer coiled-coil assemblies.

As expected, the search of the PDB identified sequences with single heptad repeats. Surprisingly, however, we also found sequences with two or more heptad repeats offset from one another. For example, in most cases the helices of coiled-coil trimers, tetramers and pentamers displayed two offset heptad repeats. These resulted in two hydrophobic seams, each of which could be assigned to a different helix–helix interface. Briefly, dimeric coiled coils have a

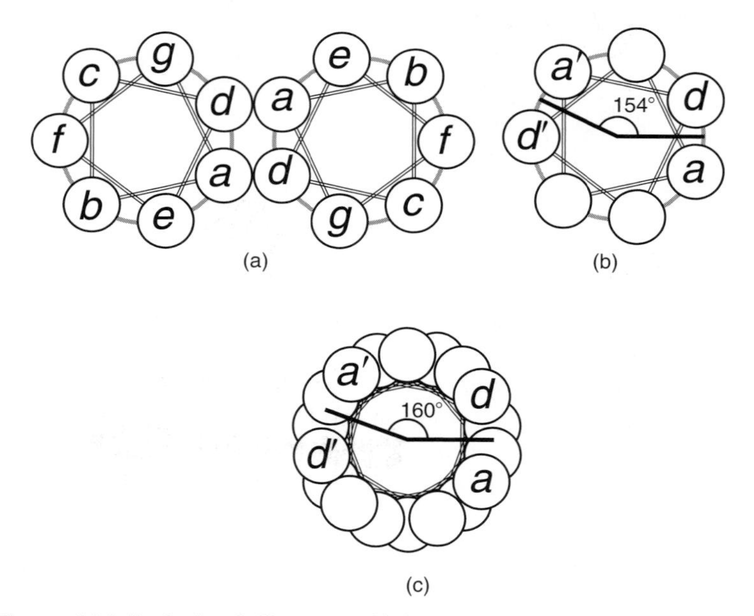

Figure 4 Helical wheel diagrams. (a) A parallel dimeric coiled coil indicating the core composed of *a* and *d* residues. This wheel assumes 3.5 residues per turn, i.e. it is in supercoil space. (b) Two overlapping heptads on a single helical wheel (*abcdefg* and *a′b′c′d′e′f′g′*). The sequence offset is 2, which results in an angular offset of 154° between the two potential core-forming interfaces. (c) As in (b), but the helix is regular with 3.6 residues per turn, i.e. it is not supercoiled. This places the two interfaces 160° apart. In all wheels the N-terminus is closest to the viewer.

single heptad repeat, which gives a single seam and one interface (Figure 4a). Trimers and higher-order structures have two repeats, both of which set up different hydrophobic seams and, in turn, assemble to give distinguishable interfaces (for example see Figure 4b). In essence, this results in assemblies with cyclical KIH interfaces (Figure 5). In addition, we observed more-complex arrangements. For instance, the core of the ectodomain of gp41 from HIV [31] has six helices. Three central (N36) helices form a parallel trimer, which is abutted on each face by another helix (C34) running antiparallel to the central helices. As a result, each central helix effectively has four interfaces (Figure 5d). We will describe these high-order coiled-coil assemblies in detail elsewhere.

For two superimposed heptads there are three possible 'sequence offsets' of 1, 2 and 3 residue(s), which are equivalent to six-, five- and four-residue offsets, respectively. For a regular 3.6-residue-per-turn α-helix, these set up two hydrophobic faces with 'angular offsets' of 100°, 160° (360°−200°) and 60° (360°−300°), respectively, around the outside of the helix. This is best seen on a

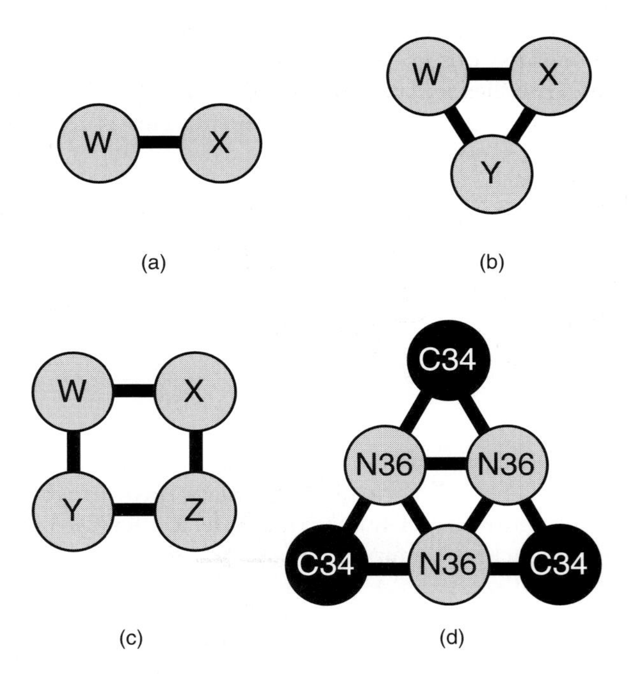

Figure 5 Schematic of inter-helical interfaces mediated by heptad repeats. (a), (b) and (c) show dimeric, trimeric, and tetrameric coiled coils, respectively. (d) The ectodomain of HIV gp41 [31]. In (a) and (c), the interfaces (represented by thick lines) between any pair of helices consist of a whole heptad; helices of the tetramer effectively have two heptads, one interacting with each partner. In (b) and (d), the bars represent 'half-heptads', i.e. all the *a* side-chains from one repeat on one helix interact with one partner, while all the *d* side chains interact with another. In all diagrams helices with the same shading are parallel, and those with different shading run antiparallel. W, X, Y, Z, N36 and C34 are either general or specific helix labels.

helical wheel (e.g. Figure 4c). If helical supercoiling is taken into account, i.e. assuming 3.5 residues per turn and using the accepted helical-wheel representation for the coiled coil (Figure 4b), these angular offsets are altered to 103°, 154° and 51°, respectively. However, both sets of angles are oversimplifications when considering helix–helix interactions in actual coiled-coil systems because side-chain size, geometry and packing also affect the helix interfaces [9,10,32]. Nonetheless, we found that many natural coiled-coil assemblies were consistent with the approximate angular offsets; trimers could be considered as having overlapping heptads separated by three residues (angular offset = 51/60°). On the other hand, tetrameric and pentameric coiled coils were often variations on a theme, with two heptad repeats offset by one residue (100/103°).

Two heptad repeats offset by two residues: α-sheet and α-cylinder constructions

Sequence offsets of two residues are potentially more interesting than the one- and three-residue offsets. This is because of the possibility of placing hydrophobic (H) residues at *a, c, d* and *f*, with *c* and *f* effectively making up the *a′* and *d′* positions of the second, offset heptad. This is represented below, where P signifies polar (non-core) residues.

```
a  b  c  d  e  f  g  a  b  c  d  e  f  g   repeat 1
H  P  P  H  P  P  P  H  P  P  H  P  P  P   binary pattern 1
P  P  H  P  P  H  P  P  P  H  P  P  H  P   binary pattern 2
f' g' a' b' c' d' e' f' g' a' b' c' d' e'  repeat 2

a  b  c  d  e  f  g  a  b  c  d  e  f  g   assigned register
H  P  H  H  P  H  P  H  P  H  H  P  H  P   overall binary pattern
```

Such sequence patterns would result in two hydrophobic seams with a wide angular-separation (154/160°), which would place them roughly on opposite sides of the helix (Figure 4b and 4c). Furthermore, it offers two possibilities for parallel helix–helix packing arrangements: *syn*, where two similar faces, i.e. *a*/*d* with *a*/*d*, or *c*/*f* with *c*/*f*, from neighbouring helices combine to produce an open α-sheet (Figure 6a); and *anti*, where *a*/*d* faces pair with *c*/*f*. In the *anti* arrangement the structure can close to form to an α-cylinder (Figure 6b). It is important to note that for antiparallel pairs of helices *syn*-typic association should lead to cylinders, whereas sheets should be formed from *anti*-typic antiparallel interfaces.

We found two three-stranded α-sheet structures in the PDB: in a domain from colicin Ia [33], and within variants of a surface glycoprotein from a trypanosome [34,35]. The recently reported structure for TolC provides the first example of an α-cylinder [36].

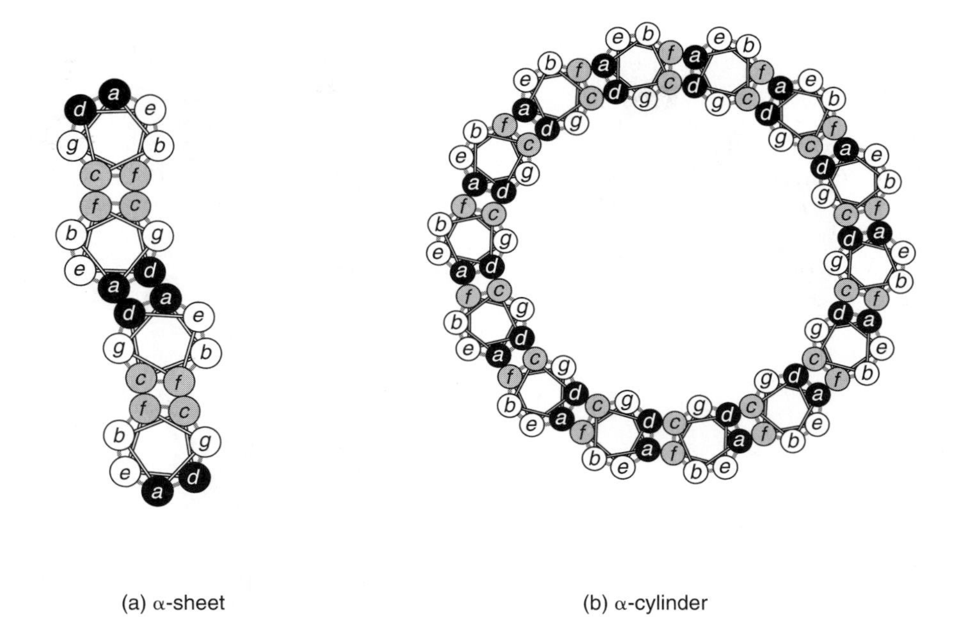

Figure 6 Two possible repeating arrangements of parallel α-helices with two heptad repeats offset by two sequence positions. (a) The *syn*-typic association leading to an α-sheet. (b) The *anti*-typic combination which gives an α-cylinder.

Natural α-sheets

We found the first examples of α-sheets in MITat 1.2 and ILTat 1.24, which are variant surface glycoproteins (VSGs) of the coat of the trypanosome that causes sleeping sickness [34,35]. These proteins have low sequence identity (approx. 20%), but have the same fold, which is a homodimer dominated by four long, antiparallel helices. The two antiparallel helices, denoted A and B, in each monomer are separated by a short loop. SOCKET revealed antiparallel KIH packing over short ranges close to the interhelical loops in both structures. In addition, the A-helix heptad exhibited a second heptad repeat offset by two residues, which interfaces with a third, parallel, helix (S) in the same chain. This structure is shown in Figure 7(a) and 7(b).

The second example of an α-sheet came from the structure of colicin Ia from *Escherichia coli* [33]. There are three marked differences between this and the VSG structures. First, helices T1, T2 and T3 of the translocation domain pack side-by-side, but in an all-antiparallel arrangement. Secondly, although it is the central helix (T2) that has two seams, and makes KIH interactions with the peripheral helices, the seams do not overlap in the sequence; the seam on one side of T2 terminates before the other begins. Finally, in the VSG structures, the 'sheets' of helices B, A and S meet by means of contacts between the

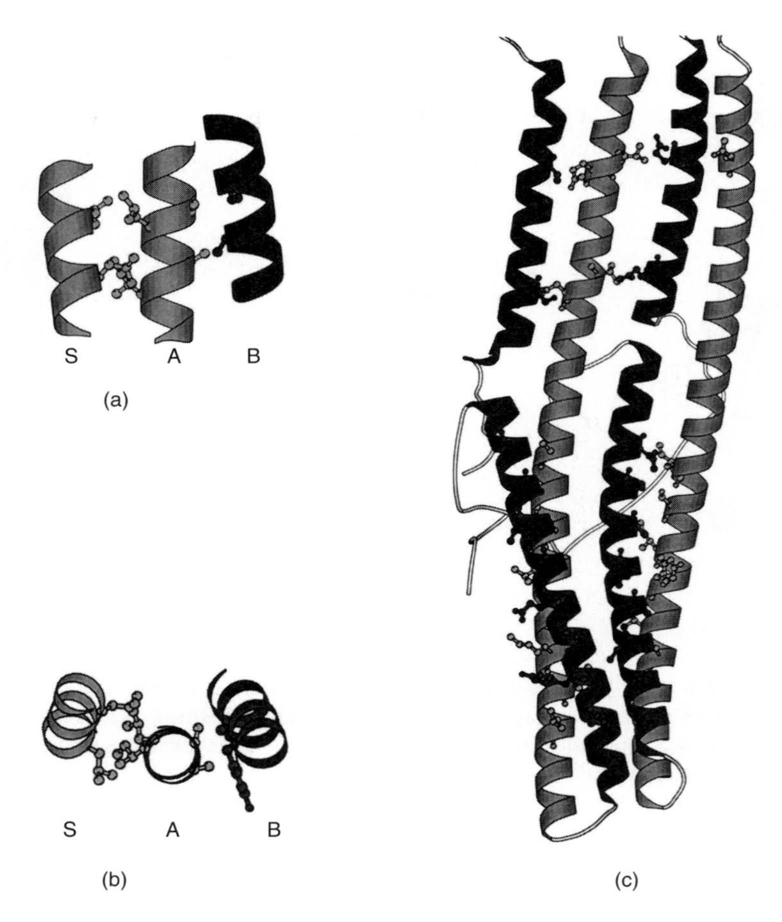

Figure 7 Observed α-sheets and α-cylinders mediated by KIH packing. (a) and (b) show orthogonal views of the S, A and B helices of the helical bundle of MITat 1.2 [34]. (c) The two helical domains (upper and lower) of a TolC monomer [36]. KIH side-chains highlighted by SOCKET are shown in ball-and-stick representation. Parallel and antiparallel helix orientations are distinguished by shading; the light helices have the N-terminal end at the top.

B and S helices of the two subunits, forming a narrow barrel-like motif. The packing between the two monomers is not KIH.

A natural α-cylinder

TolC has two α-barrel-like domains [36]. Both have 12 helices from three monomers. In the lower barrel, each helix pairs with another from the same protomer to form separate supercoiled, antiparallel coiled coils. SOCKET analysis revealed extensive antiparallel KIH interactions within these pairs, but not between them. In contrast, the helices of the upper barrel appear to pack more uniformly, albeit with a slant, giving rise to an α-cylinder. The SOCKET output for this part of the structure revealed many fewer KIH interactions than were found in the lower barrel (Figure 7c). Furthermore, KIH interactions were not contiguous around the cylinder and, in particular, they were more extensive between helices in the same monomer, but less regular between the helices abutting the monomers. In our view, the TolC barrel represents a variation of the cylinders we propose.

Nevertheless, we were able to assign heptad registers for the helices of the upper barrel unambiguously. This revealed knobs at relative *a, c, d* and *f* positions and *syn*-typic association of two seams between adjacent helices, i.e. fully consistent with the theory outlined above.

Opportunities for protein design

In summary, the TolC barrel represents a variation on the α-cylinder that we propose, and the three-helix structures from colicin Ia and VSG are the first, albeit the simplest conceivable, examples of α-sheets. These structures demonstrate that unusual coiled-coil assemblies are possible and we expect to see more natural examples.

What are the prospects for design? At least two other considerations will be necessary when exploring possibilities for peptide-design work. First, most adjacent helices in the natural α-sheets and all those in TolC are antiparallel. As indicated above, although the details of KIH packing differ between parallel and antiparallel structures, the basic principles do not. Therefore, we believe that it will be possible to construct α-sheets and α-cylinders using helices in parallel. The use of parallel helices does have one interesting consequence for the construction of α-cylinders: as the pairing in these structures is *anti*-typic, *a* residues on one helix partner *c* residues of a neighbouring helix at the same level in the structure. Similarly, *d* and *f* residues pair at the intervening levels. The result will be that successive helices will be translated up the helix and cylinder axes by two residues, which is equivalent to approx. 3 Å. Attempts to construct α-cylinders from parallel helices will give spirals of helices (Figure 8), which may or may not close. However, this is potentially extremely interesting as it opens up possibilities for making peptide-based nanotubes.

A second consideration for α-cylinder construction is the consequences of helix and coiled-coil supercoiling. The upper barrel of TolC has 12 helices. Based on a structure of parallel helices with canonical supercoiling, i.e. an angular separation of 154° between the two seams in each helix, we calculated that the cylinder should close at 14 helices. However, variations in helix number are expected. One reason for this is that helices cannot supercoil in two directions simultaneously, and some distortion is required to maintain packing at both interfaces. We found structural precedents for this in the PDB where tight KIH packing was maintained [37]. Indeed, the central helices of the three-helix α-sheets are straight (Figure 7b). (The slanting of the helices in the upper barrel of TolC may offer a compromise between straight and supercoiled helices.) Assuming the packing of completely straight helices, the angular offset becomes 160° and 18 helices would close a cylinder. However, given that, as in three-, four- and five-stranded coiled coils, side chains mediate the helix–helix

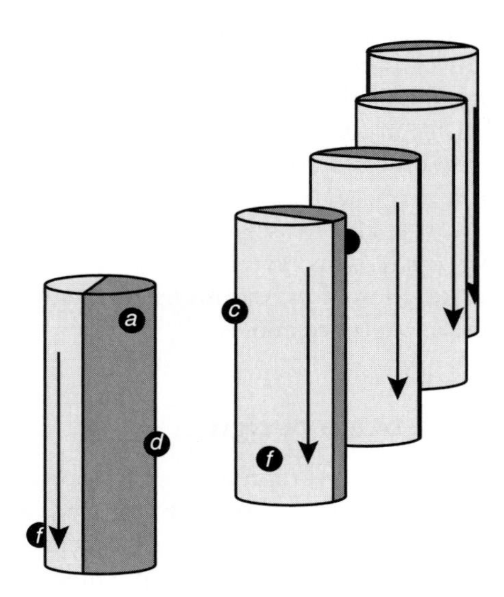

Figure 8 Cartoon showing the possible *anti*-typic association of parallel helical peptides leading to a homo-oligomeric peptide nanotube.

contact angles, other oligomerization states might be possible [9,10,32]. We calculate that small adjustments in the angular offset between 144° and 162° varies the helix number from ten to 20.

Returning to peptide nanotubes, one possibility would be to direct helix straightening by design. This could be achieved by combining the aforementioned 7- and 11-residue repeats. The effect would be to eliminate the overall hydrophobic displacement. In other words, alternating heptad and hendecad repeats give an 18-residue repeat to match the α-helical repeat; in the α-helix, 18 residues span five helical turns exactly. Incidentally, the dimer interface of GrpE provides a natural precedent for this motif [38]. Combining this with the above argument about helix translation it may be possible to create a completely closed peptide nantotube: in the parallel, straight helix case there would be 18 helices per turn of the 'cylinder', and the rise per turn is 36 residues. Thus, a 36-residue peptide with a 7–11–7–11 repeat offset by two residues should form a spiral of helices the ends of which meet to close the tube.

We are grateful to the Medical Research Council, the Biotechnology and Biological Sciences Research Council, and Active Biotech for funding this work.

References

1. Lupas, A. (1996) Trends Biochem. Sci. **21**, 375–382
2. Landschulz, W.H., Johnson, P.F. and McKnight, S.L. (1988) Science **240**, 1759–1764
3. O'Shea, E.K., Rutkowski, R. and Kim, P.S. (1989) Science **243**, 538–542
4. O'Shea, E.K., Klemm, J.D., Kim, P.S. and Alber, T. (1991) Science **254**, 539–544
5. Crick, F.H.C. (1953) Acta Crystallogr. **6**, 689–697

6. Sodek, J., Hodges, R.S., Smillie, L.B. and Jurasek, L. (1972) Proc. Natl. Acad. Sci. U.S.A. **69**, 3800–3804
7. Kohn, W.D., Mant, C.T. and Hodges, R.S. (1997) J. Biol. Chem. **272**, 2583–2586
8. Kohn, W.D. and Hodges, R.S. (1998) Trends Biotechnol. **16**, 379–389
9. Harbury, P.B., Zhang, T., Kim, P.S. and Alber, T. (1993) Science **262**, 1401–1407
10. Harbury, P.B., Kim, P.S. and Alber, T. (1994) Nature (London) **371**, 80–83
11. Woolfson, D.N. and Alber, T. (1995) Protein Sci. **4**, 1596–1607
12. Gonzalez, L., Woolfson, D.N. and Alber, T. (1996) Nat. Struct. Biol. **3**, 1011–1018
13. O'Shea, E.K., Rutkowski, R., Stafford, W.D. and Kim, P.S. (1989) Science **245**, 646–648
14. O'Shea, E.K., Rutkowski, R. and Kim, P.S. (1992) Cell **68**, 699–708
15. Vinson, C.R., Hai, J.W. and Boyd, S.M. (1993) Genes Dev. **7**, 1047–1058
16. Lavigne, P., Kondejewski, L.H., Houston, M.E., Sonnichsen, F.D., Lix, B., Sykes, B.D., Hodges, R.S. and Kay, C.M. (1995) J. Mol. Biol. **254**, 505–520
17. O'Shea, E.K., Lumb, K.J. and Kim, P.S. (1993) Curr. Biol. **3**, 658–667
18. Nautiyal, S., Woolfson, D.N., King, D.S. and Alber, T. (1995) Biochemistry **34**, 11645–11651
19. Pandya, M.J., Spooner, G.M., Sunde, M., Thorpe, J.R., Rodger, A. and Woolfson, D.N. (2000) Biochemistry **39**, 8728–8734
20. Hicks, M.R., Holberton, D.V., Kowalczyk, C. and Woolfson, D.N. (1997) Folding Des. **2**, 149–158
21. Brown, J.H., Cohen, C. and Parry, D.A.D. (1996) Proteins Struct. Funct. Genet. **26**, 134–145
22. Dure, L.I. (1993) Plant J. **3**, 363–369
23. Werner, E., Holder, A.A., Aszodi, A. and Taylor, W.R. (1996) Protein Pept. Lett. **3**, 139–145
24. Harbury, P.B., Plecs, J.J., Tidor, B., Alber, T. and Kim, P.S. (1998) Science **282**, 1462–1467
25. Chothia, C., Levitt, M. and Richardson, D. (1981) J. Mol. Biol. **145**, 215–250
26. Walther, D., Eisenhaber, F. and Argos, P. (1996) J. Mol. Biol. **255**, 536–553
27. Efimov, A.V. (1999) FEBS Lett. **463**, 3–6
28. Whitby, F.G., Kent, H., Stewart, F., Stewart, M., Xie, X.L., Hatch, V., Cohen, C. and Phillips, G.N. (1992) J. Mol. Biol. **227**, 441–452
29. Whitby, F.G. and Phillips, G.N. (2000) Proteins Struct. Funct. Genet. **38**, 49–59
30. Bernstein, F.C., Koetzle, T.F., Williams, G.J., Meyer, E.F., Brice, M.D., Rodgers, J.R., Kennard, O., Shimanouchi, T. and Tasumi, M. (1977) J. Mol. Biol. **112**, 535–542
31. Chan, D.C., Fass, D., Berger, J.M. and Kim, P.S. (1997) Cell **89**, 263–273
32. Malashkevich, V.N., Kammerer, R.A., Efimov, V.P., Schulthess, T. and Engel, J. (1996) Science **274**, 761–765
33. Wiener, M., Freymann, D., Ghosh, P. and Stroud, R.M. (1997) Nature (London) **385**, 461–464
34. Freymann, D., Down, J., Carrington, M., Roditi, I., Turner, M. and Wiley, D. (1990) J. Mol. Biol. **216**, 141–160
35. Blum, M.L., Down, J.A., Gurnett, A.M., Carrington, M., Turner, M.J. and Wiley, D.C. (1993) Nature (London) **362**, 603–609
36. Koronakis, V., Sharff, A., Koronakis, E., Luisi, B. and Hughes, C. (2000) Nature (London) **405**, 914–919
37. Walshaw, J. and Woolfson, D.N. (2001) Protein Sci. **10**, 668–673
38. Harrison, C.J., Hayer-Hartl, M., DiLiberto, M., Hartl, F.U. and Kuriyan, J. (1997) Science **276**, 431–435
39. Kraulis, P.J. (1991) J. Appl. Crystallogr. **24**, 946–950

Biochem. Soc. Symp. **68**, 125–142
(Printed in Great Britain)

9

Protease inhibitors and directed evolution: enhancing plant resistance to nematodes

Michael J. McPherson[1] and David J. Harrison

Centre for Plant Sciences, Leeds Institute for Plant Biotechnology and Agriculture, University of Leeds, Leeds LS2 9JT, U.K.

Abstract

Plant nematodes are agricultural pests, the control of which relies on chemical nematicides and fumigants that are among the most toxic and environmentally damaging of all agrochemicals. New approaches to control, based on transgenic resistance, would provide important health and environmental benefits. In this chapter we consider briefly some targets for engineering nematode resistance and discuss the use of plant protease inhibitors as anti-feedants. This approach has provided plants that display good levels of resistance against a range of nematode species. To enhance this defence strategy further we are investigating the value of directed evolution to improve the characteristics of protease inhibitors. We describe the approaches of DNA shuffling and phage display that are being used to create and screen variant libraries in the search for inhibitors with improved features.

Introduction

Nematodes represent a major drain on global agricultural productivity, resulting in losses estimated at some $100 billion per year [1]. In the U.K. and northern Europe cyst-nematodes (*Globodera* spp. and *Heterodera schachtii*) are major pests of potato and sugar beet respectively and display a highly restricted host range. In all subtropical and tropical areas root-knot nematodes (*Meloidogyne* spp.) are major pests, and display a very broad host range, often being capable of infecting and reproducing with some success on most crop species. Nematode control often relies on some integration of three principal approaches: cultural practices, use of naturally resistant varieties and chemical nematicides [2].

[1]To whom correspondence should be addressed.

Cultural measures, including crop rotation, are usually inadequate when used alone and are of little benefit to specialist growers who are restricted in their choice of crop varieties or in regions affected by broad host range *Meloidogyne* spp.

Natural resistance genes (R-genes) can offer good protection; however, for several reasons, resistant cultivars have proven successful for only a small number of crops [3,4]. One difficulty with R-genes is a consequence of their gene-for-gene interaction specificity with a particular pathogenic strain. This is highlighted by the introduction of the monogenic, natural-resistance trait H1 into potato cultivars such as Maris Piper. In England, H1 proved extremely successful in the control of the major potato cyst-nematode pest, *Globodera rostochiensis*. However, a closely-related species, *G. pallida*, that is not controlled by H1 benefited, and within a few years the population had exploded and it remains the prevalent potato cyst-nematode in England. This dramatic population switch is the direct result of the successful control of the original dominant species, *G. rostochiensis*, by an R-gene [5]. Thus, efficient control of a major pathogen within an ecological niche can liberate a minor pathogen, against which a resistance gene is ineffective. A further difficulty is the restricted plant host range to which an R-gene may be transferred and remain functional. Presumably this is a consequence of the need for the R-gene not only to interact, directly or indirectly, with an avirulence product from the pathogen, but also to integrate efficiently with appropriate cellular signalling components to elicit a hypersensitive cell destruction response.

The most effective component of a nematode control strategy is the extensive application of highly toxic and non-selective soil fumigants and nematicides, with consequent environmental and health risks. In addition to being expensive and damaging, chemical regimes are often not available to developing-world farmers and, even when available, they may not be used appropriately or safely. Soil fumigants in common use include methyl bromide and 1,3-dichloropropene (Telone II), while post-plant nematicides include oximecarbamates, such as Aldicarb, one of the most toxic and environmentally hazardous pesticides in widespread use [6]. Misuse of Aldicarb has led to severe cases of poisoning in the U.S.A. [7] and there are considerable concerns about the potential toxicological hazards of this compound. It is detectable in groundwater, its use is under review by several governments [6] and it has been withdrawn from many states of the U.S.A. Some 3% of agricultural workers are believed to suffer from nematicide poisoning each year. Class actions against manufacturers of nematicides have been filed, in American courts, by workers from Central America as a result of illness from exposure during agricultural labour [8]. The Bophal disaster in India is an example of an industrial accident resulting from production of a nematicide.

International agreements, such as The Montreal Protocol, call for the withdrawal of methyl bromide from agricultural use by 2005, as it represents a major cause of ozone-layer depletion. Unfortunately, intensive and specialized agricultural practices mean that severe crop losses would be inevitable within major sectors of the agricultural industry if chemicals were withdrawn. Despite the decision of the U.S.A. to ban the use of methyl bromide by 2002, this position has been reviewed following pressure from some crop producers who are reliant on this compound.

These issues surrounding the effective control of nematodes provide justification for the development of new crop-protection strategies, based on transgenic plant technology. Strategies should be considered with a view to reducing environmental, producer and consumer exposure and risk to chemical nematicides, while increasing economic and social benefits for producers and consumers in both the developed and developing world [9]. It is most beneficial to develop effective, durable, broad-spectrum control regimes that can be introduced readily into a range of plant backgrounds without reliance on existing cellular pathways. For the past 10 years we have developed alternative transgenic resistance strategies for control of crop plant nematode pathogens (for reviews see [10–12]). This article provides a brief, generalized description of potential targets for engineered resistance, based on the interaction of endoparasitic nematodes with the host plant root. We then describe the development of a protease inhibitor-based resistance strategy before outlining the application of directed evolution for the selection of improved efficacy inhibitors.

Targets for engineered resistance

The free-living nematode, *Caenorhabditis elegans*, grows efficiently on bacterial lawns on agar plates, thereby facilitating *in vitro* culture and genetic experiments. In contrast, plant nematodes are obligate biotrophs that can only complete their life cycle by growing on living plant tissue. In addition some nematodes, such as *Meloidogyne* spp., exhibit parthenogenesis, with no apparent role for the male in reproduction. Such features limit the extent to which experimental tools developed for *C. elegans* can be applied to plant nematodes. However, the availability of the complete genome sequence of *C. elegans* [13] and the detailed understanding that has been gained of many of its developmental and physiological processes [14,15] represents an invaluable resource for genomic-based approaches to the development of future resistance strategies.

Figure 1 shows a basic life cycle of a typical endoparasitic plant nematode, revealing several stages at which one might consider transgenic intervention [9], either by directly affecting the nematode or attenuating plant cells with which it interacts [11,16].

Invasion and migration

Eggs hatch, releasing the J2 stage that must identify and migrate towards a host plant root, by chemoreception [17]. Lectins, applied exogenously, have been shown to disrupt the invasion processes [18], suggesting that secretion into the rhizosphere of transgenically produced chemoreception interference molecules could affect the normal response of the nematode. A major benefit of such early intervention is the reduced damage that would be suffered by the plant-root system. The invasion process is damaging, not only as a result of nematode entry, migration and parasitism, but also by providing an entry route for opportunistic pathogens such as fungi and bacteria. After invasion, the J2 must migrate through the root, either between cells (root knot nematodes) or

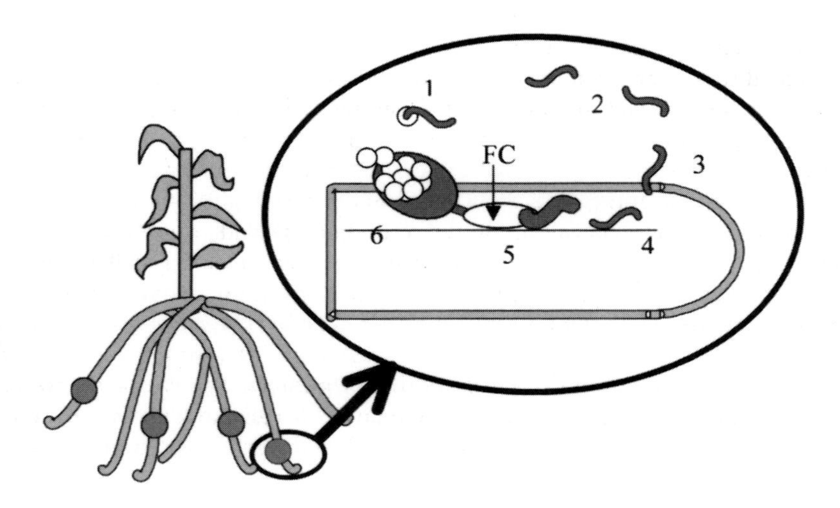

Figure 1 Simplified and generalized schematic of the life cycle of an endoparasitic nematode. (1) The J2 stage hatches from an egg in the soil that may have remained dormant for some time and (2) migrates by chemotaxis towards a plant root. (3) The animal invades the root and (4) migrates through the root either between plant cells in a non-destructive manner (root knot nematodes) or more aggressively, damaging cells (cyst nematodes). (5) The nematode then begins to feed, undergoes two further moults and becomes a sedentary female that is actively feeding. (6) The female matures and produces eggs that either remain encapsulated in her body cavity (cyst nematodes) or are extruded from the body in a gelatinous matrix (root knot nematodes). At stage (5), if the nutrient supply is insufficient due either to poor choice of feeding cell or competition, the nematode may differentiate into a male and leave the root. FC, feeding cell.

through cells (cyst nematodes), to reach a position where it can initiate formation of feeding cells. This migration phase also provides potentially useful opportunities for early intervention [9]. For example, β-1,4-endoglucanases (cellulases) are produced by nematodes and one role may be to disrupt the intercellular connections and/or integrity during migration [19,20].

Feeding cell establishment

Cyst and root-knot nematodes select one or a small number of cells from which to feed. Induction and maintenance of the feeding cells relies upon active secretion by the nematode. While these secretions remain relatively uncharacterized, there have been recent reports directed towards the analysis of the secretions [21] and of specific enzyme components, including β-1,4-endoglucanases (cellulases) [19,20] and chorismate synthase [22]. The secretions induce dramatic redifferentiation of the cell(s) to form specialized feeding cells with the characteristics of metabolically highly active transfer cells. If the site has an adequate nutrient supply then the nematode will undergo further moults from J2 to J3, and from J3 to J4 female. In contrast, a poor nutrient supply enhances differentiation to a male, and the nematode leaves the root. Females undergo muscular degeneration to a sedentary endoparasitic form. Natural-resistance genes tend to operate at this feeding-cell establishment stage leading to a hypersensitive response and atrophy of the feeding cell. Therefore, destruction or metabolic attenuation of the feeding cells represents a useful point for transgenic intervention. The most challenging aspect is the identification of a plant promoter (see for example [23]) or promoter combination that provides feeding-cell-specific expression of the effector gene(s) and, therefore, does not damage the remainder of the plant. Such an approach to feeding cell attenuation has been successfully developed in a collaboration between the University of Leeds and Advanced Technologies (Cambridge). This system is based on the expression of one- or two-component ribosome-inactivating proteins (RIPs) under the regulation of feeding-cell-specific promoters. Transgenic potato lines have been generated that show greater than 75% resistance when challenged with *G. pallida* (A. Neelam, J. Denecke, H.J. Atkinson, C. Thomas and M.J. McPherson, unpublished work).

Feeding

A useful, and presumably global, target for nematode control is the later stage of active feeding. The economically important nematodes of the genera *Meloidogyne*, *Globodera*, *Heterodera*, *Rotylenchulus*, *Nacobbus* and *Xiphinema* all generate some form of feeding cells within the host plant [24]. These cells can be used as a delivery route for anti-nematode macromolecules. However, migratory ecto- and endo-parasitic nematodes, including *Radopholus*, *Pratylenchus*, *Hirschmanniella* and *Trichodorus* do not form such cells but, rather, feed intermittently from unmodified cells [24]. Ecological niches are often inhabited by different nematodes, with one species often outcompeting others and predominating. Thus, as described previously for *G. rostochiensis* on potato, the immediate benefit of controlling the major species may be short-lived if this control allows expansion of another nematode species with a different mode of interaction with and feeding from the host plant. For several years the group at Leeds has been developing a generic anti-nematode strategy that can be effective against a wide range of nematodes, whatever their mode of feeding. Key considerations include the following: (a) the effector should target a common physiological feature required for growth and development of all nematode species, and (b) the effector molecule(s) should ideally be nematode specific, have no adverse effect on the plant and be present in all root cells from which a nematode may feed. We have focused on proteinaceous protease inhibitors delivered from promoters that will be expressed within all cells of the root system, including redifferentiated feeding cells. A further benefit of a generic anti-nematode approach is that the grower does not need to have information about the specific nematode pests, and this is of particular relevance for subsistence farmers in the developing world [9,12]. The range of root-specific promoters that have been identified will not be discussed further. The following sections deal with nematode proteases, plant protease inhibitors and engineered resistance.

Nematode proteases

Proteases are involved in many diverse processes, such as the digestion of dietary protein, in eukaryotes. Cysteine proteases had been described in animal parasitic nematodes [25] and *C. elegans* [26]. Histochemical staining of cyst-nematode cryosections with cysteine protease-specific peptide substrate (Z-Ala-Arg-Arg-4MNA) was used to localize cysteine protease activity to the intestine. Activities due to distinct enzyme types were identified in this manner through the use of specific synthetic substrates [27]. Furthermore, this approach provides a good system for monitoring the efficacy of exogenously applied inhibitors for their ability to inhibit enzyme action, and for the analysis of nematodes recovered from transgenic plants to assess the level of *in vivo* inhibition. Similarly, through the use of appropriate substrates, serine protease activity was also demonstrated, and could be inhibited by adding the inhibitor protein, cow-pea trypsin inhibitor (CpTI) [27].

Degenerate PCR primers designed to conserved sequences of proteases were used to amplify fragments of cysteine protease genes from *H. glycines*, *G. pallida*, *M. javanica* and *M. incognita* [27]. These fragments were then used to screen cDNA libraries and full-length clones of both cysteine and serine proteases were isolated from several nematode species [28,29]. These genes provide a useful resource for the expression of recombinant nematode proteases for use in selection systems for optimal inhibitor proteins, including those generated by directed evolution.

Protease inhibitors

Plants produce an array of protease inhibitors. These proteins are involved in developmental regulation and defence functions [30], although roots do not seem to be well protected by these molecules.

The serine protease inhibitor CpTI, when expressed transgenically, provides some control against insects [31]. Similarly, it has been shown that CpTI suppresses early growth and development of *G. pallida* [32,33]. However, cyst-nematodes possess a major cysteine protease [27,34]. Cysteine protease inhibitors are naturally occurring plant proteins consumed by humans in significant quantities as part of plant foods. Mammals do not use intestinal cysteine proteases and they ingest significant quantities of salivary cystatins, thereby making this class of proteases a good target from a food safety perspective. We have focused largely on developing effective nematode resistance using plant-derived cysteine protease inhibitor (phytocystatin) genes. However, we have also continued to use CpTI, particularly in combination with phytocystatins to achieve durable resistance [35].

Cystatins and oryzacystatin I

Cystatins are small (approx. 100 amino acids) protein inhibitors of cysteine proteases found in many plants [30,36], and a large number of genes (>60) have been sequenced. Phytocystatins represent a class distinct from the type I and II cystatins [37]. Owing to its availability when this work was started we focused on oryzacystatin I (OC-I) from rice seed. Although an efficient inhibitor, OC-I was significantly less potent than the animal cystatins, such as chicken egg-white cystatin (CEWC). A principal question was whether the efficacy of the plant cystatin, OC-I, could be enhanced through protein engineering. The protein was expressed in *Escherichia coli*, using the QIAexpress system (QIAGEN, Crawley), in significant quantities for biochemical experiments, bioassays and importantly for toxicological testing. Polyclonal antibodies were prepared against recombinant OC-I to allow ELISA and Western blot detection and quantification in transgenic plants [38].

Molecular modelling and mutagenesis of OC-I

We used the crystallographic structure of CEWC [39] and the complex of human stefin B with papain (EC 3.4.22.2) [40] as a basis for a three-dimensional model of OC-I. The modelled structure was subjected to energy minimization using the program Xplor [41] to ensure reasonable stereochemistry. The cystatin model was then docked into the active site of papain to provide an unrefined model for the inhibitor–protease complex [38]. In addition, alignment of 28 cysteine protease inhibitor sequences provided a basis for comparing sequence features with the K_i values for inhibitors. The structural and sequence information was used to design mutagenesis experiments [38]. The two proteases, papain and GCP-1, were used for *in vitro* assays of inhibitor variants. GCP-1 is the digestive cysteine protease from *C. elegans* [26]. Initial mutagenesis efforts focused on amino acid insertions and deletions, and while several mutations did not affect K_i, others led to an unwanted increase in K_i. In contrast, deletion of Asp-86 (OC-IΔD86) resulted in a 13-fold reduction in K_i for OC-I against both papain and GCP-1, indicating enhanced inhibitory activity. Asp-86 was substituted by 12 other amino acids to explore this site further, but all substitutions increased K_i, indicating that absence of the residue was important for increased efficacy. This study demonstrated that it was possible to improve the efficacy of a plant protease inhibitor [38].]

Testing engineered cystatins

It was shown that OC-IΔD86 was effective at inhibiting cysteine protease activity *in situ* in nematode cryosections and that it was more potent than the parental OC-I, inducing lethality in *C. elegans* feeding experiments [38]. To test efficacy against plant parasitic nematodes, a transgenic hairy-root assay was used to compare *in vivo* efficacy of OC-I with OC-IΔD86 when both were expressed at a level of approx. 0.5% of total soluble protein. Nematode growth on OC-I- and OC-IΔD86-expressing roots was reduced significantly compared with that on control roots and the effect was most pronounced on OC-IΔD86 roots. Encouragingly, for the improved inhibitor, there was little growth during the period when the female increases dramatically in size prior to egg laying.

H. schachtii and *M. incognita* growth and development were subsequently tested on *Arabidopsis thaliana* transgenic plants expressing inhibitor at approx. 0.4% of total soluble protein [42]. On control *Arabidopsis* plants most

H. schachtii females became enlarged and saccate in shape as a result of normal growth and development. In comparison, those recovered from OC-IΔD86-expressing *Arabidopsis* were significantly smaller and predominantly fusiform, with few saccate females. Of those saccate females that were found, all were smaller than the smallest recovered from a good host plant, and were below the threshold size necessary to produce eggs [42,43]. Similar effects were observed with *M. incognita*, indicating that growth of females of both species is inhibited by the cystatin [42]. Three effects were recorded: first, fewer females achieved an egg-laying size; secondly, those that did were of greatly diminished size and fecundity; and third, some saccate *H. schachtii* were developmentally compromised. Using the fluorogenic assay in nematode cryosections, protease activity was detected in saccate females recovered from control *A. thaliana* C24 but rarely in those from OC-IΔD86-expressing plants. Ingestion was further demonstrated by Western-blot detection of OC-IΔD86 in extracts of nematodes taken from the cystatin-expressing *Arabidopsis*, indicating that *H. schachtii* can ingest proteins of 11 kDa (OC-IΔD86) [42] and 20 kDa (OC-IΔD86/CpTI fusion) [35]. However, this nematode, unlike *M. incognita* cannot ingest a 28 kDa protein (green fluorescent protein) presumably reflecting a size exclusion limit of the feeding tube resulting from the molecular sieve architecture of the tube [44]. The size of molecules that can be delivered via the feeding cells for a range of nematodes is an important parameter, as it will influence decisions about the size of effector molecules that one may select for transgenic defence strategies. For a generic defence, the maximum molecular size of an effector should be based on the lowest size exclusion limit of any nematode capable of taking up the smallest protein molecules.

An active field trial programme to test transgenic potato plants is under way with experiments being conducted not only to assess nematode resistance levels, but also to assess whether there are effects on non-target organisms, including soil invertebrates and insects. These tests complement toxicology and allergenicity studies.

Improving a phytocystatin-mediated resistance approach

Two determinants will be important in governing the efficacy of a cystatin-based defence strategy. First, the promoter should provide expression throughout the root system during root aging, but not in aerial tissues. This would ensure delivery of an effective level of cystatin throughout the root to provide control against a wide range of potential nematode pests. Secondly, the inhibitory characteristics of the cystatin will be important. It could be assumed that the lower the K_i of the inhibitor, the greater will be its capacity to inhibit the target proteases and, therefore, the more effective should be the level of resistance. This remains to be definitively proven, but provides a useful working hypothesis.

In biomedical applications, it is often useful for an inhibitor to be a highly stable molecule with a long half-life. However, regulatory authorities and consumer concerns over molecules introduced for transgenic expression favour those that are less stable. This ensures that, if ingested as part of food, the inhibitors will be degraded rapidly in the digestive system, or, if released from the plant material into the rhizosphere, they will be destroyed rapidly, thereby reducing exposure of non-target soil organisms. Thus, improved inhibitors for transgenic resistance require selection for improved inhibitory properties that should not necessarily be coupled to increased stability. One approach that should prove useful for improving the inhibitory features of cystatins is the application of directed evolution, in which the encoding DNA is subjected to random mutagenesis and recombination to create a large pool of variants from which molecules with new characteristics may be isolated.

Directed evolution

In the last decade techniques such as protein display, random mutation and recombination have increasingly been applied to the evolution of protein molecules. In comparison with rational protein redesign, these new methods allow selection for molecules with improved characteristics, without the need for prior knowledge of molecular structure or even detailed mechanisms of action. Through the generation of large libraries of variants they allow selection for molecules with enhanced properties or for entirely new functions.

A successful directed-evolution experiment comprises two distinct steps. First, a population of variant molecules based on the starting molecule(s) must be created. Secondly, the library must be subjected to a selection or screening procedure that will allow the identification of the potentially very small number of molecules with the improved or altered function required.

Selection, based on a biologically expressed phenotype, is the most powerful approach for screening large numbers of variants. For example, in the selection for altered antibiotic resistance properties, *E. coli* transformants were plated on to selective agar plates containing a higher level of antibiotic, or a new antibiotic [45]. Only those cells carrying a variant conferring enhanced resistance against the antibiotic survived. In such a case, where the selection is extremely powerful and based on bacterial cell growth, the main constraint to analysing the variant library is the number of independent transformants that can be generated by transformation or electroporation. In comparison, screening methods require the assay of individual transformants for some altered phenotype. Even with the use of robotic tools, 96- or 384-well plates and automated assays, the number of individual molecules that can be assayed is of the order of 10^5. The power of the selection or screening procedure devised for a given experiment will therefore govern the number of variant molecules that can be examined. In turn, this will influence the level of mutations that are introduced during construction of the variant population and thus the method of choice for library construction.

Several techniques have been described for the creation of sequence variation from parental molecules. Error-prone PCR conditions can be adjusted to generate predictable numbers of random mutations over the length of the molecule [46]. However, directed evolution has benefited from the combination of random mutagenesis with recombination in the techniques of DNA and family shuffling developed by Stemmer and co-workers [47,48] and the staggered extension process developed by Arnold and co-workers [49]. A variety of proteins, including enzymes, signal proteins such as interleukins, photophores, cell receptors and inhibitors have been subjected to directed evolution. Selected improvements in characteristics include activity under altered conditions, such as pH or temperature, enhanced catalytic activity, altered substrate preference and increased binding affinities (for reviews see [50–52]).

Directed evolution of protease inhibitors

Nine different cysteine and serine protease inhibitors have been subjected to directed-evolution experiments [53–64]. In most cases, cassette mutagenesis of binding-loop regions was used to generate populations of variants. In a recent case, DNA shuffling has been applied to the plasminogen activation inhibitor gene to create a library of random variants [64]. In all cases, phage display of the inhibitors was used to isolate variants. The inhibitor proteins are displayed as fusion proteins with the gIIIp coat protein of the M13 capsid and therefore are located at the infective end of the phage particle. A key feature of phage display is the direct link between phenotype of the molecule displayed on the surface of the phage and genotype, since the encoding DNA is part of the packaged phage genome [65]. Biopanning is a process that exposes large numbers of displayed variants to a target molecule under conditions that select for specific interactions. Non-binding phage, that do not display an effective variant, will be washed away, whereas those with enhanced binding will remain bound to the target for longer and can be eluted as an enriched population. This sub-population is amplified and subjected to further rounds of biopanning yielding a specific population of molecules that should be more effective than the parental molecule(s). Experiments on protease inhibitors have been useful in the elucidation of the molecular basis of inhibitor specificity [54,60,63,64], the isolation of higher-affinity variants [53,55,57] and those with new specificities [56,58,59,61,62].

Directed evolution of phytocystatins

The rationale for exploring the potential of directed evolution of phytocystatins was to determine whether this approach is (a) more rapid and efficient than the rational design approach used in the case of OC-IΔD86, and (b) useful for selection of highly potent inhibitor proteins for enhanced transgenic defence. We are interested primarily in enhancing the ability of the inhibitor to remain bound to the protease in a highly stable complex. The following sections outline the approach we have adopted to the directed evolution of phytocystatins.

Strategy for directed evolution of phytocystatins

Phage-display offers a convenient approach for screening new variant inhibitors as it is possible to screen libraries of up to 10^9 variants. To generate a library of phytocystatin variants, the family shuffling method [48] provides a useful tool. In this approach, multiple homologous parent genes, that display some level of variation, provide the starting point for DNA shuffling to create new variants by a process involving accumulation of new point mutations and both intra- and inter-gene recombination. This approach can lead to potentially very large and complex variant libraries. To confirm the utility of these approaches to phytocystatins we have demonstrated the ability to display phytocystatins as gIIIp fusions on M13 and have optimized DNA shuffling methods for the short phytocystatin genes that are only about 300 bp. In addition we developed binding assays for selection and used surface plasmon resonance (SPR) for kinetic analysis of individual molecules.

Phage display of phytocystatins

The gene (*oc-I*) encoding OC-I was cloned into the phagemid vector pDHisII, a modified form of pHen, in which a 6-His tag region has been introduced between the insert and the gene III. The construct was sequenced to confirm the integrity of the gene fusion. Phagemid particles were produced by superinfecting *E. coli* TG1(pDHisII/*oc-I*) with helper phage. The display of OC-I on the surface of phage was confirmed by Western-blot analysis of phage protein probed with anti-OC-I antiserum (Figure 2).

The detection of displayed OC-I, while encouraging, does not provide evidence for display of functional molecules. Such evidence requires the demonstration of functional interaction with a cysteine protease. For preliminary experiments, the readily available and well characterized cysteine protease, papain, has been used as the target for molecular interaction studies and biopan-

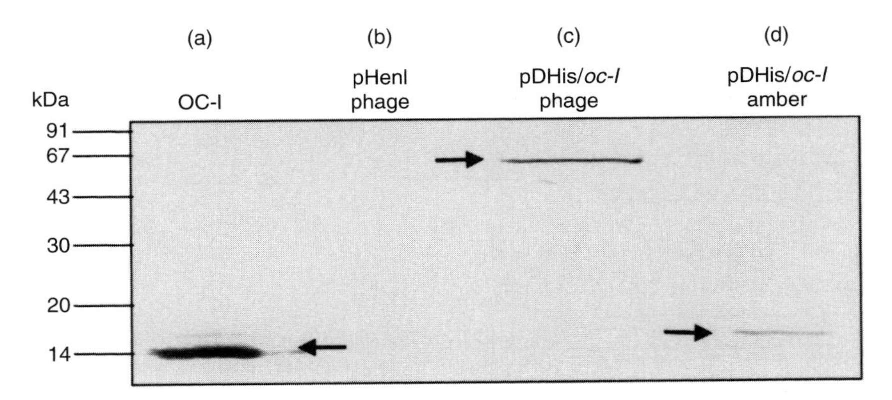

Figure 2 Western-blot analysis of an SDS/PAGE using an anti-OC-I monoclonal antibody. Recombinant OC-I proteins were purified by exploiting their 6-His tag using nickel affinity chromatography from the following samples: (a) *E. coli* M15/pQE30*oc-I*, revealing purified OC-I; (b) 2.5×10^{11} plaque-forming units (p.f.u.) of pHen1, the parental phagemid that does not carry the *oc-I* coding region, showing no detectable protein; (c) 2.5×10^{11} p.f.u. pDHis/*oc-I* phage showing an OC-I–gIIIp fusion protein indicative of successful phage display of OC-I; (iv) *E. coli* HB2151, an amber non-suppressing strain carrying pDHis/*oc-I*, resulting in the production of soluble OC-I rather than gIII fusion protein. Molecular marker sizes are shown on the left and arrows indicate the positions of the OC-I protein cross-reactivity.

ning. Display of functional OC-I molecules was demonstrated by the retention of such phage on immobilized papain relative to wild-type phage. In the context of screening a library of variants, the population will be enriched for the tightest-binding phage by a combination of stringent binding and washing conditions, elution time and target concentration.

Generating a variant library by DNA shuffling

The process of DNA shuffling [45,47] uses a method for random fragmentation, such as sonication or controlled digestion with DNase I, to generate fragments from the starting DNA molecule(s). A schematic representation of the process is shown in Figure 3. Fragments within a restricted size range, often 50–100 bp, are then purified from the gel. The shuffling process takes the form of a PCR for some 40–80 cycles, but in the absence of flanking primers. At the annealing step, the various single-stranded fragments can anneal with com-

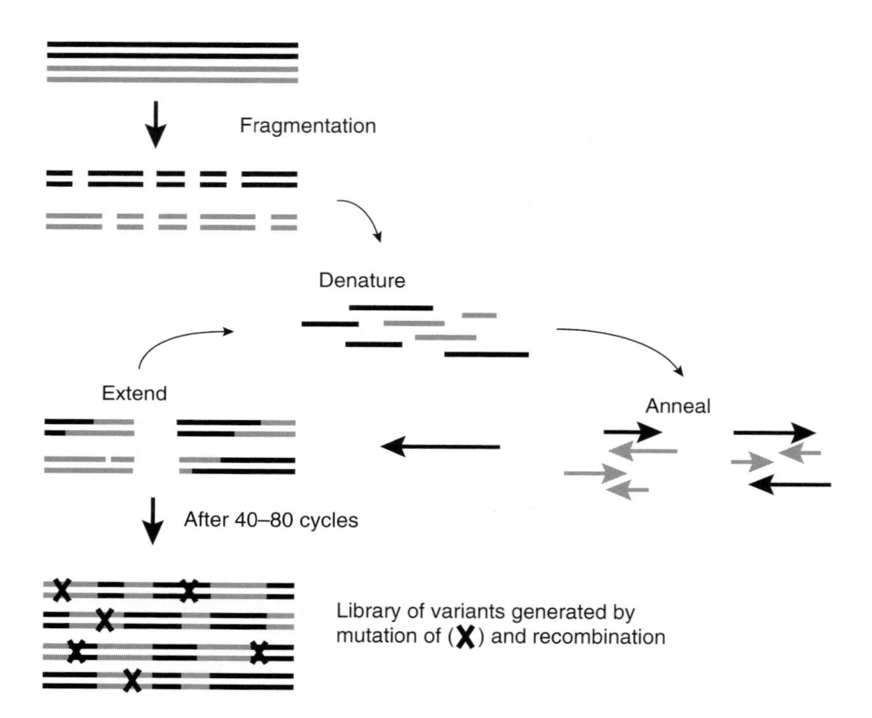

Figure 3 Schematic representation of steps in a DNA shuffling experiment. For simplicity, only two molecules are shown. The molecules are fragmented and small fragments, typically 50–100 bp in length, are isolated. These are subjected to multiple cycles (40–80) of PCR in the absence of added primers. This leads to annealing of fragments and self-priming. The conditions favour incorporation of mutations, and at each cycle template switching leads to recombination and the spread of mutations throughout the population. Eventually, full-length molecules are regenerated and can be amplified by PCR with flanking primers for cloning and selection or screening.

plementary strands leading to single strand overhangs. Self-priming leads to these overhangs being filled-in during the synthesis phase of the PCR, so that each of the original single strands increases in length. At each cycle of PCR the DNA strands will 'template switch', i.e. a given strand will anneal with a different template strand, resulting in recombination. In addition, the PCR conditions favour incorporation of mutations that then become fixed in sub-populations of molecules as a consequence of the recombination process. Eventually, full-length products are formed and these can then be amplified preferentially from an aliquot of the reaction by performing PCR in the presence of flanking primers.

The resulting full-length product band comprises a range of variant DNA molecules generated by the mutation/recombination process. These fragments can be cloned into an appropriate expression vector and transformed into competent cells for the subsequent selection/screening process to identify desirable variants. Usually, the selected variant(s) are then subjected to further rounds of shuffling to recombine the beneficial mutations and allow identification of improved variants.

A variation of this method, called family shuffling, allowing the generation of more diverse libraries that explore greater sequence space, has been devised [48]. This is achieved by shuffling homologous genes that encode the same protein from different sources, and therefore display sequence variation.

Family shuffling of phytocystatins

As a basis for our proposed family shuffling of phytocystatins, an early objective was to demonstrate that, starting with small genes (approx. 300 bp), it was possible to achieve appropriate mutation and recombination rates. Initially, we chose to use two phytocystatin genes, *oc-I* (rice) and *zm* (maize), that have 74% nucleotide sequence identity. The genes were amplified from the pDHisII vectors using flanking primers, resulting in initial PCR products of 550 bp (Figure 4a). These products were subjected to DNase I digestion and gave rise to a smear on an agarose gel (Figure 4b). The gel region containing fragments of between 25 and 50 bp was excised, the DNA isolated and subjected to 60 cycles of DNA shuffling with no added primers. This resulted in a smear of products (Figure 4c), and an aliquot was subjected to PCR in the presence of flanking primers, yielding product of the expected size for regenerated products (Figure 4d). These products were cloned into pDHisII to create a library of *oc-I/zm* shuffled variants.

To identify whether recombination events had occurred between fragments from the two genes, we devised a PCR-based screen. As shown in Figure 5, PCR products were generated from the library for all primer combinations, indicating the presence of clones representing recombination events between the *oc-I* and *zm* coding regions. Subsequent DNA sequence analysis of random clones from the library revealed a point mutation rate of 1%, with a recombination rate of 2.5 events per 300 bp gene. These results indicate that the approach of DNA shuffling can be used with small genes and that it should be possible to generate a more complex library by shuffling further phytocystatin genes.

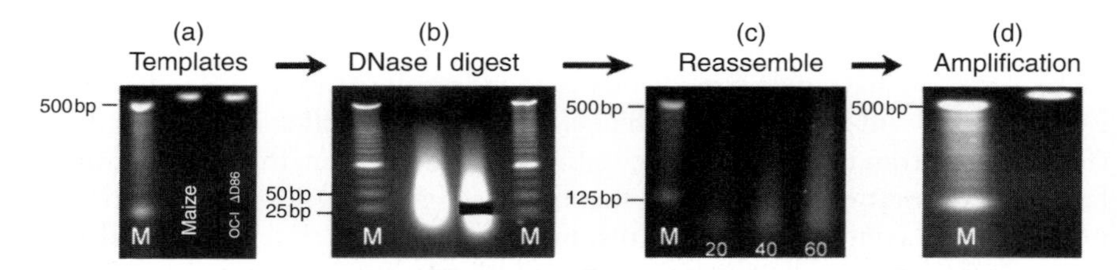

Figure 4 DNA shuffling of maize (*zm*) and rice (*oc-I*) phytocystatin DNA fragments. (a) Original 550 bp *zm* and *oc-I* products amplified from pDHisII clones. The amplified fragments comprise approx. 300 bp of coding DNA with 120 bp of vector DNA at each end. (b) DNase I digestion pattern of the products from (a) showing the 25–50 bp fragment region removed from gel. (c) Reassembly of fragments by thermal cycling (95 °C, 30 s; 55 °C, 30 s, 72 °C, 30 s) for 20, 40 and 60 cycles; >500 bp fragments are visible at 60 cycles. (d) Final PCR amplification for 20 cycles, with flanking primers added, of an aliquot from the 60-cycle sample shown in (c) and revealing a 550 bp product band. M indicates molecular size markers.

Choosing a subset of phytocystatin genes for DNA shuffling

There are now more than 60 sequences for phytocystatins available within databases. We have selected a subset of 11 genes representative of the family and have resynthesized these genes with *E. coli* codon usage to facilitate good expression of each phytocystatin in *E. coli* and to reduce the level of nucleotide sequence divergence between the gene sequences. These genes have been used to generate a shuffled library that is being screened against papain, in the first instance.

SPR studies

Kinetic characterization of the phytocystatins has been undertaken by using the technique of SPR, which yields data for both binding and dissociation of a complex formed between a protease and phytocystatin. We have obtained good kinetic data for the bimolecular interactions of phytocystatins with papain, and clear comparative results when different phytocystatins are tested in parallel. As an example, Figure 6 shows the profiles for two phytocystatins immobilized via their 6-His tags on to Ni^{2+}-BIAcore sensor chips (Pharmacia, Uppsala) and then challenged with papain. This technique will be useful for the detailed analysis of variants selected from the shuffled library.

Perspective and future studies

Phytocystatins are cysteine protease inhibitors that are widely expressed plant proteins and are consumed in large quantities in foods. A phytocystatin, OC-I, has been improved by rational design, yielding OC-IΔD86 that exhibits a 13-fold decrease in K_i. The transgenic control of major groups of agronomically important nematode parasites using protease inhibitors has been achieved

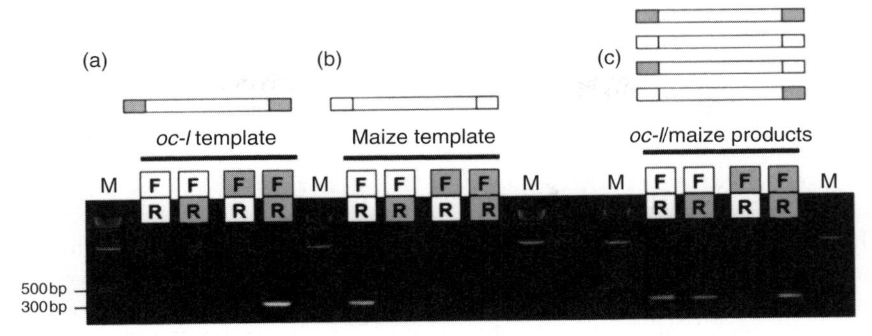

Figure 5. Confirmation that recombination events have occurred between *oc-IΔD86* and *zm* templates during the experiment shown in Figure 3. PCR was performed using combinations of *oc-I* and *zm* specific forward (F) and reverse (R) primer combinations with (a) *oc-IΔD86* template DNA, (b) *zm* template DNA or (c) aliquots of the reassembled products of the DNA shuffling between the two genes from Figure 3(d). The specificity of the *oc-I* and *zm* primers for their respective templates are shown in (a) and (b), where products are amplified only when the appropriate gene primers are present. In contrast, for the reassembled genes, amplification has occurred for all four primer combinations, indicating that products are present in which there has been at least one recombination event between the *oc-I* and *zm* coding regions. M, molecular size markers.

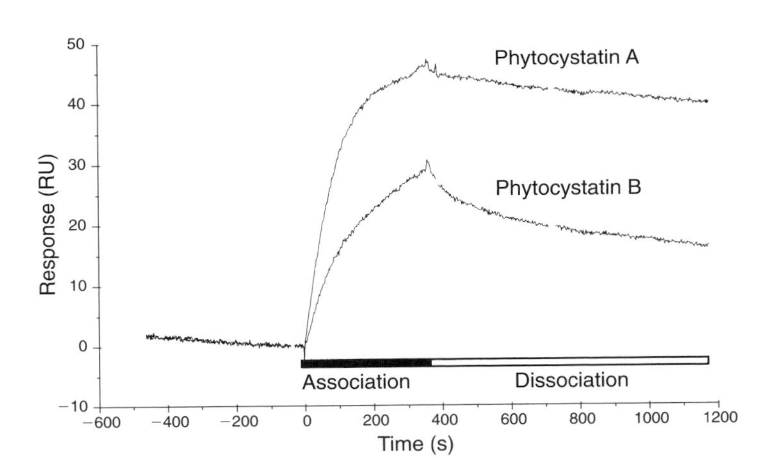

Figure 6 Analysis of the binding kinetics of two phytocystatins using SPR. Phytocystatins were tethered to the SPR chip surface via their 6-His tags by nickel affinity chelation. Papain was injected across the sensor chip surface for 6 min (association phase) and then running buffer was injected for a further 15 min (dissociation phase). The data fit well to the expected 1:1 binding model and reveal the kinetic constants K_D, K_{on} and K_{off}.

by delivery to the feeding nematode via plant cells. The defence is now being tested in crop plants to evaluate its efficacy. A wide range of studies are under way to evaluate the safety of phytocystatins, including (a) toxicological and allergenicity studies of phytocystatins, (b) assessment of agronomic traits such as yield and vigour of transgenic plants expressing phytocystatins in their roots, and (c) environmental impact studies on non-target invertebrates. It is also important to test whether nematodes can adapt to exposure to protease inhibitors by altering the range of protease genes they express, as has been shown to occur for insects [66]. In this regard, durable plant defence is likely to involve gene-pyramiding [67], and we have successfully tested dual protease inhibitor constructs [35].

Directed evolution offers the prospect of selecting new and improved phytocystatins with enhanced inhibitory properties. While papain has proved to be useful in the identification of improved cystatins, such as OC-IΔD86, it is of plant origin. It will be important to utilize nematode proteases for some selection experiments and for determination of kinetic parameters. We have cloned several protease genes [27–29] and these provide a useful resource for expression experiments and comparative studies on the efficacy of selected variant phytocystatins. In the longer term, directed evolution offers tremendous potential for the rapid development of new proteins, based on existing plant defence molecules, to enhance the armoury of crop plants in their battles with plant pathogens and pests.

We are grateful for financial support from the Biotechnology and Biological Sciences Research Council (BBSRC), the Overseas Development Administration, the Scottish Office Agriculture, Environment and Fisheries Department, Novartis AG, Zeneca, Advanced Technologies (Cambridge), Nickerson BIOCEM, The Hawaiian Pineapple Growers Association and University of Hawaii. D.J.H. was in receipt of a BBSRC studentship. We thank Howard Atkinson and all members of the Leeds Plant Nematode Laboratory, past and present, for support and helpful discussions.

References

1. Sasser, J.N. and Freckman, D.W. (1987) in Vistas on Nematology (Veech, J.A. and Dickerson, D.W., eds.), pp. 7–14, Society of Nematologists, Hyattsville, MD
2. Hague, N.G.H. and Gowen, S.R. (1987) in Principles and Practice of Nematode Control in Crops (Brown, R.H. and Kerry, B.R., eds), pp. 131–178, Academic Press, Sydney
3. Cook, R. and Evans, K. (1987) in Principles and Practice of Nematode Control in Crops (Brown, R.H. and Kerry, B.R., eds), pp. 179–231, Academic Press, Sydney
4. Roberts, P.A. (1992) J. Nematol. **24**, 213–227
5. Trudgill, D.L. (1991) Annu. Rev. Phytopathol. **29**, 167–192
6. Gustafson, D.I. (1993) Pesticides in Drinking Water, Chapel Hill, NC
7. Goldman, L.R., Smith, D.F., Neutra, R.R., Saunders, L.D., Pond, E.M., Stratton, J., Waller, K., Jackson, R.J. and Kizer, K.W. (1990) Arch. Environ. Health **45**, 229–236
8. Anon. (1992) in Agrow, vol. 171, pp. 21–22, PJB Publications, Surrey, U.K.
9. Atkinson, H.J., Urwin, P.E., Hansen, E. and McPherson, M.J. (1995) Trends Biotechnol. **13**, 369–374

10. McPherson, M.J., Urwin, P.E., Lilley, C.J. and Atkinson, H.J. (1997) in Cellular and Molecular Basis for Plant–Nematode Interactions (Fenoll, C., Ohl, S. and Grundler, F., eds), pp. 237–249, Kluwer, Dordrecht
11. Atkinson, H.J., Lilley, C.J., Urwin, P.E. and McPherson, M.J. (1998) in The Physiology and Biochemistry of Free-Living and Plant-Parasitic Nematodes (Perry, R. and Wright, D. J., eds), pp. 381–413, CAB International, Oxford
12. Atkinson, H.J., Urwin, P.E., Lilley, C.J. and McPherson, M.J. (1998) In Potato Cyst Nematodes (Marks, J. and Brodie, W., eds), pp. 209–236, CAB International, Oxford
13. The *C. elegans* Sequencing Consortium. (1998) Science **282**, 2012–2018
14. Wood, W.B. (1988) The Nematode *Caenorhabditis elegans*, Cold Spring Harbor Laboratory, Cold Spring Harbor, NY
15. Riddle, D.L., Blumenthal, T., Meyer, B.J. and Priess, J.R. (1997) *C. elegans* II, Cold Spring Harbor Laboratory, Cold Spring Harbor, NY
16. Burrows, P.R. and De Waele, D. (1997) in Cellular and Molecular Aspects of Plant–Nematode Interactions (Fenoll, C., Grundler, F.M.W. and Ohl, S.A., eds), pp. 217–236, Kluwer, Dordrecht
17. Zuckerman, B.M. and Jansson, H.B. (1984) Annu. Rev. Phytopathol. **22**, 95–113
18. Marban-Mendoza, N., Jeyaprakash, A., Jansson, H.B., Damon, R.A. and Zuckerman, B.M. (1987) J. Nematol. **19**, 331–335
19. Smant, G., Stokkermans, J., Yan, Y.T., de Boer, J.M., Baum, T.J., Wang, X.H., Hussey, R.S., Gommers, F.J., Henrissat, B., Davis, E.L., Helder, J., Schots, A. and Bakker, J. (1998) Proc. Natl. Acad. Sci. U.S.A. **95**, 4906–4911
20. Bera-Maillet, C., Arthaud, L., Abad, P. and Rosso, M.N. (2000) Eur. J. Biochem. **267**, 3255–3263
21. Duncan, L.H., Robertson, L., Robertson, W.M. and Kusel, J.R. (1997) Parasitology **115**, 429–438
22. Lambert, K.N., Allen, K.D. and Sussex, I.M. (1999) Mol. Plant–Microbe Interact. **12**, 328–336
23. Barthels, N., van der Lee, F.M., Klap, J., Goddijn, O.J.M., Karimi, M., Puzio, P., Grundler, F.M.W., Ohl, S.A., Lindsey, K., Robertson, L. et al. (1997) Plant Cell **9**, 2119–2134
24. Sijmons, P.C., Atkinson, H.J. and Wyss, U. (1994) Annu. Rev. Phytopathol. **32**, 235–259
25. Cox, G.N., Pratt, D., Hageman, R. and Boisvenue, R.J. (1990) Mol. Biochem. Parasitol. **41**, 25–34
26. Ray, C. and McKerrow, J.H. (1992) Mol. Biochem. Parasitol. **51**, 239–249
27. Lilley, C.J., Urwin, P.E., McPherson, M.J. and Atkinson, H.J. (1996) Parasitology **113**, 415–424
28. Urwin, P.E., Lilley, C.J., McPherson, M.J. and Atkinson, H.J. (1997) Parasitology **114**, 605–613
29. Lilley, C.J., Urwin, P.E., Atkinson, H.J. and McPherson, M.J. (1997) Mol. Biochem. Parasitol. **89**, 195–207
30. Ryan, C.A. (1990) Annu. Rev. Phytopathol. **28**, 425–449
31. Hilder, V.A., Gatehouse, A.M.R., Sheerman, S.E., Barker, R.F. and Boulter, D. (1987) Nature (London) **330**, 160–163
32. Atkinson, H.J. (1993) In Opportunities for Molecular Biology in Crop Protection (Beadle, D.J., ed.), pp. 257–266, British Crop Protection Council, Farnham
33. Hepher, A. and Atkinson, H.J. (1996) Patent number WO 9215690
34. Koritsas, V.M. and Atkinson, H.J. (1994) Parasitology **109**, 357–365
35. Urwin, P.E., McPherson, M.J. and Atkinson, H.J. (1998) Planta **204**, 472–479
36. Richardson, M. (1991) Methods Plant Biochem. **5**, 259–305
37. Kondo, H., Abe, K., Emori, Y. and Arai, S. (1991) FEBS Lett. **278**, 87–90
38. Urwin, P.E., Atkinson, H.J., Waller, D.A. and McPherson, M.J. (1995) Plant J. **8**, 121–131

39. Bode, W., Engh, R., Musil, D., Thiele, U., Huber, R., Karshikov, A., Brzin, J., Kos, J. and Turk, V. (1988) EMBO J. **7**, 2593–2599
40. Stubbs, M.T., Laber, B., Bode, W., Huber, R., Jerala, R., Lenarcic, B. and Turk, V. (1990) EMBO J. **9**, 1939–1947
41. Brunger, A.T., Kuriyan, J. and Karplus, M. (1987) Science **235**, 458–460
42. Urwin, P.E., Lilley, C.J., McPherson, M.J. and Atkinson, H.J. (1997) Plant J. **12**, 455–461
43. Atkinson, H.J., Urwin, P.E., Clarke, M.C. and McPherson, M.J. (1996) J. Nematol. **28**, 209–215
44. Urwin, P.E., Moller, S.G., Lilley, C.J., McPherson, M.J. and Atkinson, H.J. (1997) Mol. Plant–Microbe Interact. **10**, 394–400
45. Stemmer, W.P.C. (1994) Nature (London) **370**, 389–391
46. Cadwell, R.C. and Joyce, G.F. (1994) PCR Methods Appl. **3**, S136–S140
47. Stemmer, W.P.C. (1994) Proc. Natl. Acad. Sci. U.S.A. **91**, 10747–10751
48. Crameri, A., Raillard, S.A., Bermudez, E. and Stemmer, W.P.C. (1998) Nature (London) **391**, 288–291
49. Zhao, H.M., Giver, L., Shao, Z.X., Affholter, J.A. and Arnold, F.H. (1998) Nat. Biotechnol. **16**, 258–261
50. Shao, Z.X. and Arnold, F.H. (1996) Curr. Opin. Struct. Biol. **6**, 513–518
51. Minshull, J. and Stemmer, W.P.C. (1999) Curr. Opin. Chem. Biol. **3**, 284–290
52. Petrounia, I.P. and Arnold, F.H. (2000) Curr. Opin. Biotechnol. **11**, 325–330
53. Roberts, B.L., Markland, W., Ley, A.C., Kent, R.B., White, D.W., Guterman, S.K. and Ladner, R.C. (1992) Proc. Natl. Acad. Sci. U.S.A. **89**, 2429–2433
54. Jongsma, M.A., Bakker, P.L., Stiekema, W.J. and Bosch, D. (1995) Mol. Breed. **1**, 181–191
55. Dennis, M.S., Herzka, A. and Lazarus, R.A. (1995) J. Biol. Chem. **270**, 25411–25417
56. Wang, C.I., Yang, Q. and Craik, C.S. (1995) J. Biol. Chem. **270**, 12250–12256
57. Markland, W., Ley, A.C. and Ladner, R.C. (1996) Biochemistry **35**, 8058–8067
58. Markland, W., Ley, A.C., Lee, S.W. and Ladner, R.C. (1996) Biochemistry **35**, 8045–8057
59. Rottgen, P. and Collins, J. (1995) Gene **164**, 243–250
60. Beekwilder, J., Rakonjac, J., Jongsma, M. and Bosch, D. (1999) Gene **228**, 23–31
61. Kiczak, L., Koscielska, K., Otlewski, J., Czerwinski, M. and Dadlez, M. (1999) Biol. Chem. **380**, 101–105
62. Tanaka, A.S., Silva, M.M., Torquato, R.J.S., Noguti, M.A.E., Sampaio, C.A.M., Fritz, H. and Auerswald, E.A. (1999) FEBS Lett. **458**, 11–16
63. Ylinenjarvi, K., Widersten, M. and Bjork, I. (1999) Eur. J. Biochem. **261**, 682–688
64. Stoop, A.A., Jespers, L., Lasters, I., Eldering, E. and Pannekoek, H. (2000) J. Mol. Biol. **301**, 1135–1147
65. Smith, G.P. (1985) Science **228**, 1315–1317
66. Jongsma, M.A., Bakker, P.L., Peters, J., Bosch, D. and Stiekema, W.J. (1995) Proc. Natl. Acad. Sci. U.S.A. **92**, 8041–8045
67. Gatehouse, A.M.R., Shi, Y., Powell, K.S., Brough, C., Hilder, V.A., Hamilton, W.D.O., Newell, C.A., Merryweather, A., Boulter, D. and Gatehouse, J.A. (1993) Philos. Trans. R. Soc. London Ser. B **342**, 279–286

Biochem. Soc. Symp. **68**, 143–153
(Printed in Great Britain)

10

Degradation of explosives by nitrate ester reductases

Richard E. Williams*, Deborah A. Rathbone*, Peter C.E. Moody†, Nigel S. Scrutton† and Neil C. Bruce*[1]

*Institute of Biotechnology, University of Cambridge, Tennis Court Road, Cambridge CB2 1QT, U.K., and †Department of Biochemistry, University of Leicester, Adrian Building, University Road, Leicester LE1 7RU, U.K.

Abstract

Explosive-contaminated land poses a hazard both to the environment and to human health. Microbial enzymes, either in their native or heterologous hosts, are a powerful and low-cost tool for eliminating this environmental hazard. As many explosives have only been present in the environment for 10 years, and with similar molecules not known in Nature, the origin of enzymes specialized for the breakdown of explosives is of particular interest. Screening of environmental isolates resulted in the discovery of flavoproteins capable of denitrating the explosives pentaerythritol tetranitrate (PETN) and glycerol trinitrate. These nitrate ester reductases are related in sequence and structure to Old Yellow Enzyme from *Saccharomyces carlsbergenisis.* All the members of this family have α/β barrel structures and FMN as a prosthetic group, and reduce various electrophilic substrates. The nitrate ester reductases are, however, unusual in that they display activity towards the highly recalcitrant, aromatic explosive 2,4,6-trinitrotoluene, via a reductive pathway resulting in nitrogen liberation. We have embarked on a detailed study of the structure and mechanism of PETN reductase from a strain of *Enterobacter cloacae*. Work is focused currently on relating structure and function within this growing family of enzymes, with a view to engineering novel enzymes exhibiting useful characteristics.

Introduction

Explosives can be broadly classified into three groups: nitroaromatics [e.g. trinitrotoluene (TNT)], nitramines (e.g. hexahydro-1,3,5-trinitro-1,3,5-

[1]To whom correspondence should be addressed.

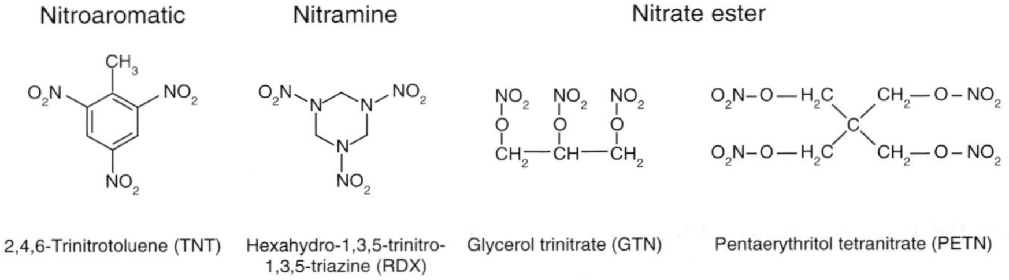

Nitroaromatic | Nitramine | Nitrate ester

2,4,6-Trinitrotoluene (TNT) | Hexahydro-1,3,5-trinitro-1,3,5-triazine (RDX) | Glycerol trinitrate (GTN) | Pentaerythritol tetranitrate (PETN)

Figure 1 Chemical structures of some important explosives.

triazine) and nitrate esters [e.g. glycerol trinitrate (GTN) and pentaerythritol tetranitrate (PETN)] (Figure 1). The manufacture, storage and disposal of explosive compounds has resulted in extensive contamination of the environment; this is a matter of increasing concern, since many of these compounds are both toxic and highly recalcitrant. GTN is toxic to aquatic organisms, with an LC_{50} of between 1.5 and 3 mg/l [1]. TNT is particularly toxic; in the first year of the First World War 475 munitions workers died in the U.S.A. due to poisoning from TNT exposure [1]. Although TNT has not been manufactured in the U.S.A. or Europe for approx. 20 years, particular environmental problems still persist from TNT manufacture during the Second World War.

These energetic compounds are new to Nature, having been introduced to the environment by man in the last 100 years. The origin of mechanisms used by micro-organisms to tolerate, or even derive nutrition from these compounds is, therefore, of fundamental interest. Characterization of such breakdown mechanisms also offers the prospect of utilizing either the organisms or their component enzymes in the identification and bioremediation of contaminated land. With an understanding of the molecular basis for the reactions, the ability to engineer these mechanisms to improve their efficiency or alter their specificity will undoubtedly follow.

Nitrate ester reductases

Very similar enzymes have been identified in all bacteria in which the degradation of nitrate ester explosives has been characterized in detail. The enzymes catalyse the nicotinamide-cofactor-dependent reductive cleavage of nitrate esters to give the corresponding alcohol and nitrite. PETN reductase from *Enterobacter cloacae* PB2 is a monomeric protein of approx. 40 kDa, which requires NADPH for activity [2]. This enzyme sequentially removes two of the four nitro groups of PETN, and two of the three groups of GTN. Similar enzymes were found to be responsible for the nitrate ester degrading activity in *Agrobacterium radiobacter* [3] ('nitrate ester reductase'), and in strains of *Pseudomonas fluorescens* and *P. putida* [4] ('xenobiotic reductases'). All utilize a non-covalently bound FMN as a redox cofactor. The *Agrobacterium* enzyme differs from the others in requiring NADH as the source of reducing equivalents. The genes that encode the nitrate ester reductases from *E. cloacae*, *A. radiobacter*, *P. putida* and *P. fluorescens* have been cloned [3–5] and they display high sequence identity, constituting a new family of related enzymes. The characteristics of these enzymes set them clearly apart from glutathione- and iron-dependent enzymes implicated in mammalian and fungal transformation of nitrate esters.

With the growth in genomic sequencing projects, the existence of many uncharacterized relatives of the nitrate ester reductases is becoming apparent. Basic local alignment searches of the protein sequence of PETN reductase against completed and unfinished microbial genomes [6] reveal several related enzymes in Gram-negative prokaryotes (Figure 2), and there are several yeast and plant homologues. There appears to be a strong level of sequence conservation across the family in Gram-negative bacteria, yeast and plants, although several bacterial genomes that have been completely sequenced lack an orthologue. Alignment of deduced protein sequences from genes in the family shows, as expected, that residues intimately involved in FMN binding and hydrogen bonding in the active site are well conserved. The true physiological role of these enzymes is at present unknown. The wide distribution of closely

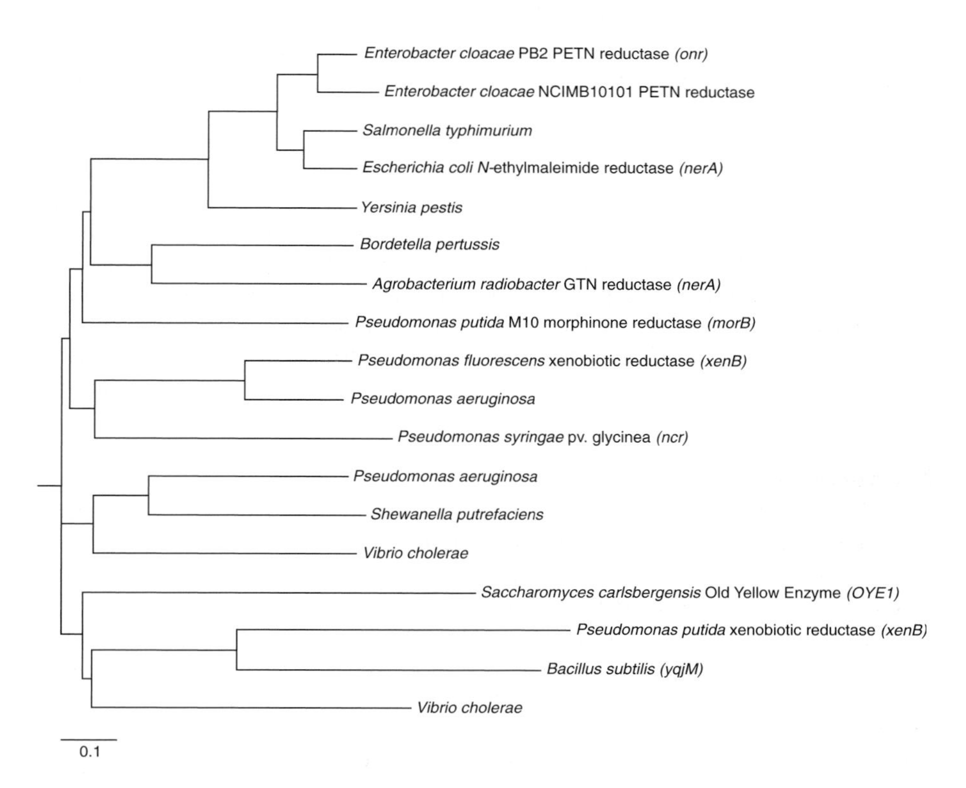

Figure 2 Phylogenetic relationship of prokaryotic nitrate ester reductases and related enzymes. The phylogeny is derived from the alignment of full-length protein sequences. Where no enzyme name is shown, the protein represented has not been characterized, and is the conceptual translation of genomic DNA sequence.

related enzymes suggests that a conserved function exists; a role in oxidative-stress response [7] or fatty-acid metabolism is possible.

Biological transformation of TNT

The microbial degradation of nitroaromatic compounds has been an area of intense research, and has been reviewed extensively [8–10]. Transformation of these compounds falls into two basic categories: fortuitous transformation by anaerobic bacteria and fungi, mainly to the amino derivatives, and aerobic bacterial degradation, which permits the utilization of nitroaromatics as both carbon and nitrogen sources for growth. Favoured strategies to release the nitro group include mono-oxygenase-catalysed cleavage to yield a hydroxyl group and nitrite, and dioxygenase-catalysed cleavage to yield a diol and nitrite. However, in the case of TNT, significant electron-withdrawal by the three nitro groups renders electrophilic attack of the aromatic ring much harder to achieve, and no oxygenases have yet been discovered that display activity towards TNT.

A wide variety of enzymes, in both prokaryotic and eukaryotic cells, can catalyse the reduction of the nitro group of nitroaromatic compounds. In the majority of these enzymes, nitroreductase activity is probably not the physiological role of the enzyme — the nitro group readily acts an electron acceptor. Enzymes acting as nitroreductases have been classified according to their ability to catalyse nitroreduction in the presence of oxygen [11]. Type I (oxygen-insensitive) nitroreductases reduce the nitro group in two-electron increments, while Type II (oxygen-sensitive) nitroreductases catalyse one-electron reduction to form a nitro radical, which rapidly reacts with oxygen to re-form the nitro group, preventing any net transformation in the presence of oxygen. While there appear to be few features shared by these two classes of enzymes, it has become apparent that many enteric bacteria possess closely related Type I nitroreductases. In particular, enzymes from *Enterobacter cloacae* [12,13] and *Escherichia coli* [14] have been cloned and characterized. Under aerobic conditions, trinitrotoluene is generally reduced to produce aminodinitrotoluenes or, occasionally, diaminonitrotoluenes. Anaerobic transformation of TNT can result in the slow formation of triaminotoluene. These compounds are not believed to be subject to further metabolism; their probable fate is polymerization to form amide, imine, or azo/azoxy polymers [15], or adsorption/irreversible binding to soil [16,17].

An unusual feature of trinitroaromatic compounds is their susceptibility to direct reductive attack of the aromatic ring. This was reported initially in a strain of *Rhodococcus erythropolis* able to use picric acid (2,4,6-trinitrophenol) as a source of nitrogen [18]. Hydride addition to picric acid yielded a hydride-Meisenheimer complex, which was further converted into 2,4-dinitrophenol with the liberation of nitrite. Recent characterization of a similar activity in a strain of *Nocardiodes simplex* has demonstrated the involvement of a coenzyme F420-dependent hydride transferase [19]. An analogous transformation of TNT to hydride-Meisenheimer and dihydride-Meisenheimer complexes has been characterized in *Rhodococcus* and *Mycobacterium* spp. [11].

Transformation of TNT by nitrate ester reductases

While nitrate ester reductases have been isolated on the basis of their ability to liberate nitrite from PETN and GTN, it is apparent from the study of *E. cloacae* PETN reductase that they might also share an ability to reduce TNT. *E. cloacae* PB2 was observed to grow very slowly on TNT as a sole nitrogen source in mineral media salts with glucose as the sole carbon source [20]. To determine whether PETN reductase might play a role in the degradation of TNT by *E. cloacae* PB2, PETN reductase was added to reaction mixtures containing TNT and NADPH. The reaction mixtures developed an orange colour with an absorbance maximum at approx. 500 nm, suggesting the formation of products in which aromaticity has been disrupted (Figure 3). No such coloured products were generated in the absence of enzyme, TNT or NADPH.

The hydride-Meisenheimer complex of TNT has a characteristic absorption spectrum which does not match that observed in the reduction of TNT by PETN reductase. However, repeated experiments have shown that a reducing agent, sodium borohydride, reduces TNT initially to the hydride-Meisenheimer complex, and subsequently to one or more orange products with absorption spectra similar to that observed with PETN reductase [20]. Reaction mixtures resulting from the reduction of TNT by PETN reductase/ NADPH or sodium borohydride were analysed by HPLC. Products were detected at 260 nm (UV) and 500 nm (visible). Absorbance spectra of each peak were also measured. Chemical reduction of TNT resulted in a peak with the characteristic spectrum of the hydride-Meisenheimer complex, and at least five other visible peaks. In enzymic reaction mixtures, two visible peaks appeared, matching the chemical peaks in retention times and absorption spectra. Chemical reduction experiments suggest that the initial product is the hydride-Meisenheimer complex, with further reduction leading to the other products observed.

Time-course analysis of the reduction of TNT by PETN reductase/ NADPH has confirmed that the hydride-Meisenheimer complex is present transiently, prior to the accumulation of orange products similar to those reported in transformations of TNT by *Rhodococcus* and *Mycobacterium* spp. [11].

Nitrite is formed by the action of PETN reductase on TNT [20]. In one reaction mixture incubated for a period of several days with an enzymic NADPH regeneration system to regenerate NADPH oxidized by PETN reductase, it was found that 1.0 mol nitrite per mol of TNT initially present had been liberated. The stoichiometry and timing of nitrite release have yet to be determined. Nitrite may be released by slow, non-enzymic breakdown of one or more of the reaction products.

Many flavin-containing enzymes act as nitroreductases, and hydroxylamino- and amino-dinitrotoluenes were produced from TNT both by PETN reductase and cloned *E. cloacae* nitroreductase [13]. Therefore, PETN reductase transforms TNT by two separate routes, reducing either the nitro group or the aromatic ring (Figure 3). The inability of *E. cloacae* nitroreductase to liber-

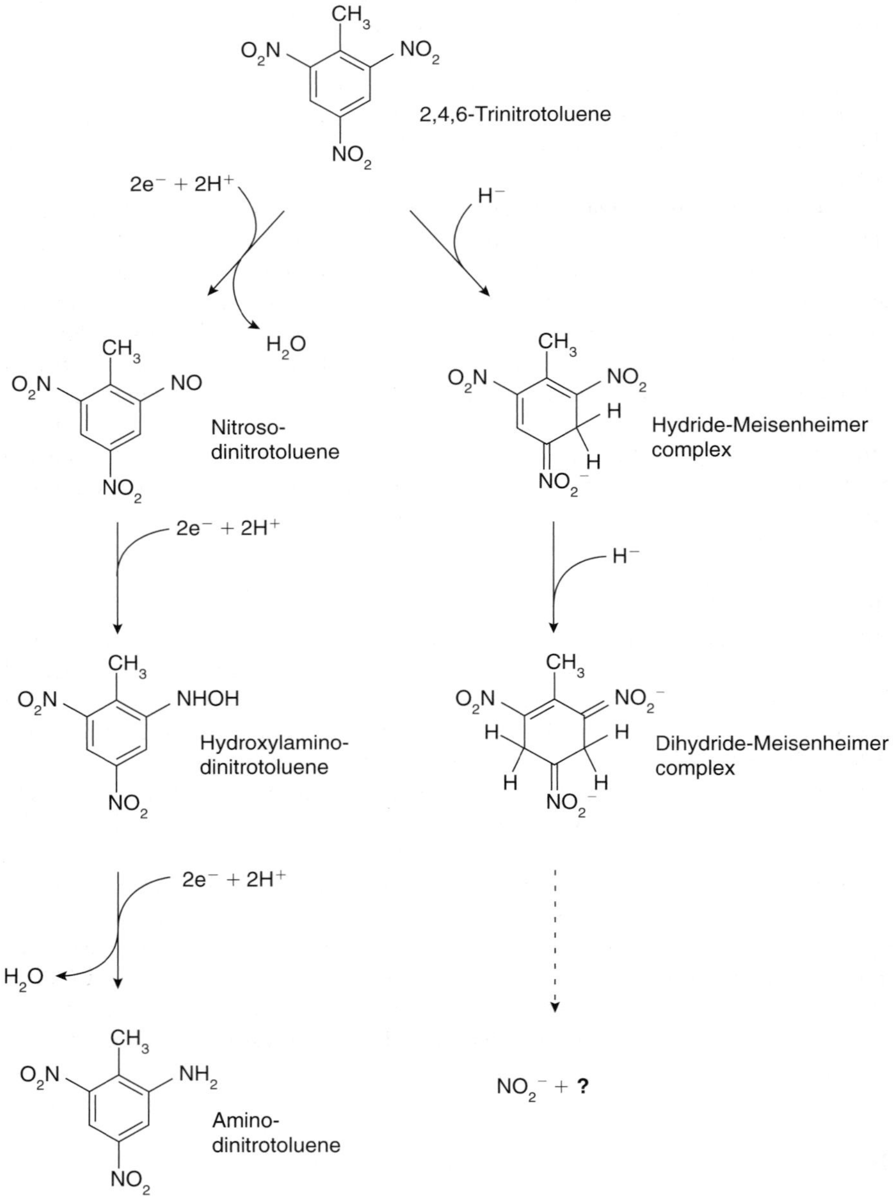

Figure 3 Products of TNT metabolism by nitrate ester reductases. Each step involves the oxidation of NADPH to $NADP^+$. Nitroso-dinitrotoluene has not been directly observed.

ate nitrite from TNT [20] suggests that nitrite is more likely to be a product of the hydride addition pathway.

Since the final reaction products of TNT reduction by PETN reductase contain less nitrogen than TNT, and appear to be water-soluble and non-aromatic, they are likely to be less toxic and less recalcitrant than TNT or the products of nitroreductase transformation of TNT. Therefore, nitrate ester reductase enzymes with enhanced hydride-transferase activity may be useful in the remediation of TNT-contaminated soil and water. To this end, transgenic plants expressing PETN reductase have been investigated [21].

In order to investigate the structural basis for the TNT hydride-transferase activity of PETN reductase, four related enzymes were studied to make a structural and functional comparison. The criteria of a published gene sequence and genetic diversity were used to select enzymes. On this basis, PETN reductase from the type strain of *E. cloacae*, *Escherichia coli N*-ethylmaleimide reductase [22], *P. putida* M10 morphinone reductase [23], and *Saccharomyces carlsbergensis* Old Yellow Enzyme (OYE) [24] were chosen. The pairwise similarity of the enzyme amino-acid sequences ranges from 96 to 44%. Each of the enzymes in the group has been reported on the basis of their ability to catalyse a particular, unusual, reaction of interest. The activity of individual enzymes against the characteristic substrates of other group members has not been investigated previously.

While all the enzymes studied showed some slow activity against TNT, the nature of the products seemed to vary, with only the close relatives of PETN reductase yielding the orange-coloured products indicative of reduction of the aromatic ring. The nature of the activity against TNT was investigated further. To attempt a quantitative analysis of metabolite formation, and to assess the order in which metabolites were formed, incubations of enzyme with TNT were sampled repeatedly and analysed by ion-pair HPLC. TNT was consumed completely; the major intermediate metabolites produced were hydroxylamino-dinitrotoluene and dihydride-Meisenheimer complexes. The ratio of metabolites produced varied considerably across the five enzymes, with the *E. cloacae* and *E. coli* enzymes producing less of the hydroxylamino products. With the analytical techniques used, only amino-dinitrotoluene is seen to accumulate, yet this accounts for a maximum of 4% of the mass balance. The formation of other, polar, metabolites was observed by HPLC, but identification and quantification have not yet proved possible. From the UV/visible spectrum, it appears that dihydroxylamino-nitrotoluene may be formed in preference to amino-dinitrotoluene.

Role of key active-site residues

A crystallographic structure of OYE has been published [25] and is shown in Figure 4. Partially refined co-ordinates from crystallography of PB2 PETN reductase [26,27] and morphinone reductase [27,28] are available. The *E. cloacae* NCIMB10101 and *E. coli* enzymes are closely related to PB2 PETN reductase, and homology modelling of these structures using the co-ordinates of PB2 PETN reductase as a template yielded plausible results. The active-site

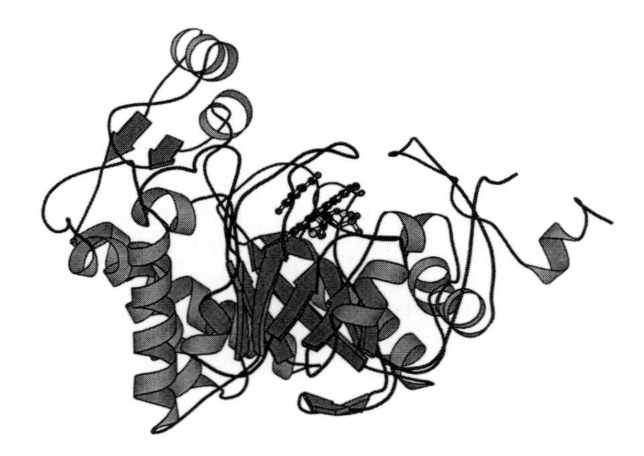

Figure 4 Ribbon diagram of the structure of OYE, with *p*-hydroxy-benzaldehyde bound. Figure drawn using MOLSCRIPT [32].

structures of all five enzymes are very similar, and the roles of certain residues in the OYE active site are quite well understood. In the crystal structure of OYE in complex with *p*-hydroxybenzaldehyde, the phenolate oxygen forms two hydrogen bonds to His-191 and Asn-194, displacing a chloride ion bound in the empty oxidized enzyme structure (Figure 5). Anchored in this position, the phenolic ring forms a π–π stacking interaction with the flavin. The inhibitor β-estradiol binds in a similar manner. If the α,β-unsaturated carbonyl substrate 2-cyclohexenone were to bind with the oxygen bound by His-191 and Asn-194, the substrate's β-carbon would be aligned above the flavin N5 atom. This correlates with previous data suggesting hydride transfer from the flavin N5 atom to the β-carbon. Site-directed mutagenesis has implicated Tyr-196 in transfer of a proton to the α-carbon [29], completing the saturation of the double bond.

The residues critical for activity in OYE are present in all the nitrate ester reductases, although Asn-194 is substituted by histidine in PETN reductase and close relatives. Given this high degree of similarity, and given that PETN reductase is also active against cyclohexenone, it is reasonable to expect that the basic mechanism of OYE is also applicable to hydride transfer reactions catalysed by nitrate ester reductases. OYE saturates nitrocyclohexene [29], and therefore a nitro group will substitute for a carbonyl or phenolate functionality as a ligand for the His–Asn pair. It is, therefore, not hard to rationalize the binding of nitrate esters to nitrate ester reductases. There is no equivalent of the olefinic β-carbon in nitrate ester substrates, and it is not yet clear whether hydride transfer is involved in the reductive denitration of nitrate esters. It might well be the case that nitrate ester reductases can catalyse both hydride transfer and sequential electron transfer reactions, with nitrate ester reduction falling into the latter category.

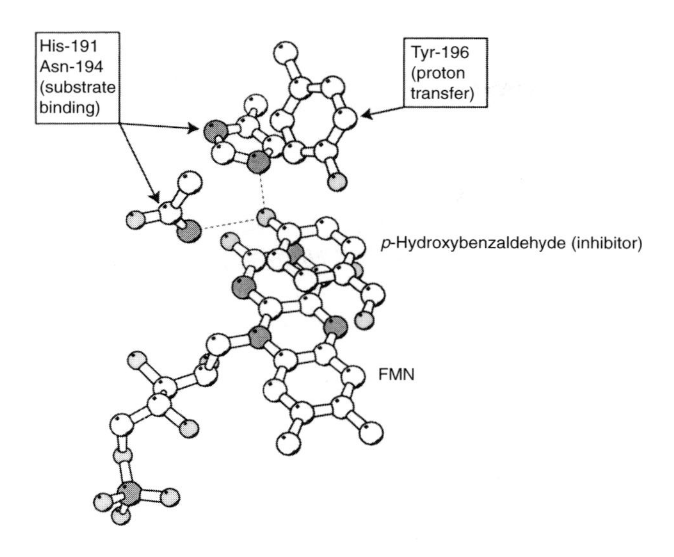

Figure 5 Active site of OYE. Figure drawn using MOLSCRIPT [32].

Nitrate ester reductases as templates for engineered enzymes

Proteins with an α/β-barrel fold, such as OYE and the nitrate ester reductases, form attractive targets for protein engineering. The combined structural information implies that the active sites of the enzymes are formed by a small number of residues from loops at the top of the barrel. Structural variation in these loops is unlikely to affect the stability of the enzyme as a whole, suggesting that engineering of altered or enhanced specificity may be feasible. Indeed, a recent paper has demonstrated the feasibility of both designed and random mutagenesis of such loop regions in *E. coli* indole-3-glycerol-phosphate synthase [30].

The substrate binding site in the nitrate ester reductases is relatively small, and as much of its base is contributed by the flavin molecule, there are a limited number of candidates for site-directed mutagenesis around the active site — certainly less than ten. Of these, three appear to be potentially involved in hydrogen bonding. The others seem to provide a generally amorphous hydrophobic environment, with little obvious basis for specific recognition, which is consistent with the wide range of ligands shown to bind OYE and homologues.

The substitution of His (186 in PETN reductase) for Asn (194 in OYE) in PETN reductase and close relatives may affect the strength of hydrogen-bonding interactions with the substrate molecule. This effect has been explored in OYE by assessing binding affinity of phenols with a range of pK_a values [31], with the conclusion that Asn-191/His-194 and Asn-191/Asn-194 mutants showed considerably weaker hydrogen-bonding potential. As hydrogen bonding with substrate oxygen is involved in activating the β-position for hydride acceptance by α,β-unsaturated carbonyl substrates, the difference in hydrogen-bonding residues in those enzymes with TNT hydride transferase activity could be significant.

Tyr-196 of OYE has been shown to be necessary for activity against substrates requiring concerted addition of a proton at the α-position, with hydride at the β-position [29]. While this residue is conserved in 39 out of 46 members of the OYE family, in morphinone reductase tyrosine is substituted by cysteine. The fact that morphinone reductase is the only enzyme with detectable codeinone reductase activity could be a coincidence, but the lower pK_a or smaller size of cysteine might be significant. As codeinone is a comparatively large substrate, substrate binding interactions irrelevant for smaller substrates might also be important.

Metabolism of TNT by the OYE family proceeds via two pathways, with the relative fluxes varying between enzymes. This variation suggests that the balance of products formed may be amenable to control by protein engineering. The absence of significant hydride transferase activity towards TNT in OYE and morphinone reductase suggests that this unusual activity is confined to close relatives of PETN reductase. Because OYE reduces many of its known products by hydride transfer, the lack of significant activity against TNT is at first sight surprising. It is possible that PETN reductase binds TNT in a manner which activates the 3-position for hydride acceptance. Alternatively, variation in the levels of competing nitroreductase activity may account for the difference between the enzymes.

The combination of structural and functional information now available for the nitrate ester reductases makes the rational redesign of these enzymes a realistic proposition. The role of key active-site residues in TNT reduction is being explored by site-directed mutagenesis. With the structural information available, it is possible to predict regions of the protein structure responsible for dictating substrate specificity. Libraries of genes mutated in these regions are being screened for altered substrate specificity, using a colorimetric assay for nitrite release.

This work was funded by the Biotechnology and Biological Sciences Research Council, DERA and the Ministry of Defence.

References

1. Yinon, J. (1990) Toxicity and Metabolism of Explosives, CRC Press, Boca Raton, FL
2. Binks, P.R., French, C.E., Nicklin, S. and Bruce, N.C. (1996) Appl. Environ. Microbiol. **62**, 1214–1219
3. Snape, J.R., Walkley, N.A., Morby, A.P., Nicklin, S. and White, G.F. (1997) J. Bacteriol. **179**, 7796–7802
4. Blehert, D.S., Fox, B.G. and Chambliss, G.H. (1999) J. Bacteriol. **181**, 6254–6263
5. French, C.E., Nicklin, S. and Bruce, N.C. (1996) J. Bacteriol. **178**, 6623–6627
6. http://www.ncbi.nlm.nih.gov/BLAST/unfinishedgenome.html
7. Brown, B.J. and Massey, V. (1999) In Flavins and Flavoproteins 1999: Proceedings of the 13th International Symposium, (Ghisla, S., Kroneck, P., Macheroux, P. and Sund, H., eds), pp. 659–662, Rudolf Weber, Berlin
8. Spain, J.C. (1995) Annu. Rev. Microbiol. **49**, 523–555

9. Gorontzy, T., Drzyzga, O., Kahl, M.W., Bruns-Nagel, D., Breitung, J., Vonloew, E. and Blotevogel, K.H. (1994) Crit. Rev. Microbiol. **20**, 265–284
10. Spain, J.C., ed. (1998) Biodegradation of Nitroaromatic Compounds, Plenum Press, New York
11. Cerniglia, C.E. and Somerville, C.C. (1998) In Biodegradation of Nitroaromatic Compounds (Spain, J.C., ed.), pp. 99–115, Plenum Press, New York
12. Bryant, C. and DeLuca, M. (1991) J. Biol. Chem. **266**, 4119–4125
13. Bryant, C., Hubbard, L. and McElroy, W.D. (1991) J. Biol. Chem. **266**, 4126–4130
14. Zenno, S., Koike, H., Kumar, A.N., Jayaraman, R., Tanokura, M. and Saigo, K. (1996) J. Bacteriol. **178**, 4508–4514
15. Lewis, T.A., Ederer, M.M., Crawford, R.L. and Crawford, D.L. (1997) J. Ind. Microbiol. Biotechnol. **18**, 89–96
16. Rieger, P.-G. and Knackmuss, H.J. (1995) in Biodegradation of Nitroaromatic Compounds (Spain, J.C., ed.), pp. 1–18, Plenum Press, New York
17. Achtnich, C., Sieglen, U., Knackmuss, H.J. and Lenke, H. (1999) Environ. Toxicol. Chem. **18**, 2416–2423
18. Lenke, H. and Knackmuss, H.J. (1992) Appl. Environ. Microbiol. **58**, 2933–2937
19. Ebert, S., Rieger, P.-G. and Knackmuss, H.-J. (1999) J. Bacteriol. **181**, 2669–2674
20. French, C.E., Nicklin, S. and Bruce, N.C. (1998) Appl. Environ. Microbiol. **64**, 2864–2868
21. French, C.E., Rosser, S.J., Davies, G.J., Nicklin, S. and Bruce, N.C. (1999) Nat. Biotechnol. **17**, 491–494
22. Miura, K., Tomioka, Y., Suzuki, H., Yonezawa, M., Hishinuma, T. and Mizugaki, M. (1997) Biol. Pharm. Bull. **20**, 110–112
23. French, C.E. and Bruce, N.C. (1995) Biochem. J. **312**, 671–678
24. Saito, K., Thiele, D.J., Davio, M., Lockridge, O. and Massey, V. (1991) J. Biol. Chem. **266**, 20720–20724
25. Fox, K.M. and Karplus, P.A. (1994) Structure **2**, 1089–1105
26. Moody, P.C.E., Shikotra, N., French, C.E., Bruce, N.C. and Scrutton, N.S. (1998) Acta Crystallogr., Sect D: Biol. Crystallogr. **54**, 675–677
27. Barna, T., Moody, P.C.E., Craig, D.H., Scrutton, N.S. and Bruce, N.C. (1999) in Flavins and Flavoproteins 1999: Proceedings of the 13th International Symposium, (Ghisla, S., Kroneck, P., Macheroux, P. and Sund, H., eds) pp. 671–674, Rudolf Weber, Berlin
28. Moody, P.C.E., Shikotra, N., French, C.E., Bruce, N.C. and Scrutton, N.S. (1997) Acta Crystallogr. Sect. D Biol. Crystallogr. **53**, 619–621
29. Kohli, R.M. and Massey, V. (1998) J. Biol. Chem. **273**, 32763–32770
30. Altamirano, M.M., Blackburn, J.M., Aguayo, C. and Fersht, A.R. (2000) Nature (London) **403**, 617–622
31. Brown, B.J., Deng, Z., Karplus, P.A. and Massey, V. (1998) J. Biol. Chem. **273**, 32753–32762
32. Kraulis, P.J. (1991) J. Appl. Crystallogr. **24**, 946–949

Subject index